PARSONS BRINCKERHOFF

THE FIRST
100 YEARS

PARSONS BRINCKERHOFF
THE FIRST
100 YEARS

Benson Bobrick

VNR VAN NOSTRAND REINHOLD COMPANY
New York

Copyright © 1985 by Parsons Brinckerhoff

Library of Congress Catalog Card Number: 84-19526
ISBN: 0-442-27264-2
ISBN: 0-442-27263-4 pbk.

Printed in Canada

Published by Van Nostrand Reinhold Company Inc.
135 West 50th Street
New York, New York 10020

Van Nostrand Reinhold Company Limited
Molly Millars Lane
Wokingham, Berkshire RG11 2PY, England

Van Nostrand Reinhold
480 Latrobe Street
Melbourne, Victoria 3000, Australia

Macmillan of Canada
Division of Gage Publishing Limited
164 Commander Boulevard
Agincourt, Ontario M1S 3C7, Canada

15   14   13   12   11   10   9   8   7   6   5   4   3   2   1

Library of Congress Cataloging in Publication Data
Main entry under title:

Parsons Brinckerhoff

   Includes index.
   1. Parsons Brinckerhoff.
TA217.P33P37   1985   338.7'62'0097471   84-19526
ISBN 0-442-27264-2
ISBN 0-442-27263-4 (pbk.)

In Memory of My
Father
Robert L. Bobrick
(1902-1983)

When we mean to build,
We first survey the plot, then draw the model;
And when we see the figure of the house,
Then must we rate the cost of the erection;
Which if we find outweighs ability,
What do we then but draw anew the model
In fewer offices, or at last desist
To build at all? Much more in this great work, —
Which is almost to pluck a kingdom down
And set another up, — should we survey
The plot of situation and the model,
Consent upon a sure foundation,
Question surveyors, know our own estate,
How able such a work to undergo,
To weigh against his opposite; or else,
We fortify in paper, and in figures,
Using the names of men instead of men:
Like one that draws the model of a house
Beyond his power to build it; who, half through,
Gives o'er and leaves his part-created cost
A naked subject to the weeping clouds,
And waste for churlish winter's tyranny.

Shakespeare, *Henry IV,* Pt. 2, Act I, Sc. 3, ll. 41-62

FOREWORD

Although Minerva sprang fully formed from the head of Jupiter, engineering was an evolutionary process. With roots in the craft tradition, engineering, when combined with scientific principles, became a creative art form. During the 19th century, this tradition was completely remolded to meet the demands of a worldwide period of rapid industrialization and urbanization. By the end of the 19th century, civil engineering was established as a profession and the practice of engineering assumed its present organizational form and approach to the conception, design, and construction of engineering works on an unprecedented scale.

From the first days of the American Republic, builders responded to the growing needs of a new nation bent on internal improvements. Bridges, factories, and a whole range of engineering structures were erected by builders, usually without the benefit of architects, using traditional methods and materials. As the 19th century advanced, iron was increasingly used in structures and, with its use, the field of building was transformed from a craft tradition to one firmly in the hands of engineers using the principles of mechanics in the design and analysis of engineering structures. Until mid-century, the use of iron in buildings was for producing fireproof structures. William Strickland, architect and engineer, first used iron for columns in the U.S. Bank Building in Philadelphia. By the end of the century, iron was used as the basis for skyscrapers, one of the great achievements of American civil engineers.

With the advent of railways and the rapid increase in locomotive weights, iron was increasingly favored for long-span railway bridges. It also provided the tendons for suspension bridges which now could be built with spans exceeding 1,000 feet. It was the heroic age of civil engineering with leading engineers such as Roebling, Whipple, Long, Ellet, Bollman, and Fink acting as designers, constructors, and financiers for the erection of large bridges. There were others in related engineering enterprises who shared the public acclaim for engineering works in an age of progress. In his series on *Civilisation*, Kenneth

Clark refers to this new breed of engineers as the great field marshals of the industrial revolution.

Following the American Civil War, with structural steel available in quantity and at a reasonable price, a host of fabricating shops sprang up across the country. They were primarily in the bridge business, but fabricated other structures from wrought iron and, later, mild steel. This became the era of catalogue bridges and other diverse machinery, obtainable from "friendly" local salesmen. The market was quite uncontrolled, and in the absence of building codes and material specifications both public and private clients had little basis for judging one catalogue bridge from another. Little wonder that engineers such as Waddell condemned the system and sought its replacement with one where a consulting engineer would represent the client and prepare all of the necessary design contracts, specifications, and drawings on his behalf. In addition, the consulting engineer, in most cases, supervised the construction work of the firm that had secured the building contract in competitive bidding with other firms.

Thus, the engineering consulting firm was born. This new system generally replaced the standardized designs with custom-designed structures to suit the individual needs of clients. Not only were "off-the-peg" structures replaced by bespoke "tailor-made" structures, but the earlier antebellum engineer/entrepreneur faded from the scene, to be replaced by a team of professional consulting engineers.

One of the earliest of these firms was founded by William Barclay Parsons, Jr. in 1885. This firm has now been in existence for a century and has been a leader in civil engineering with many well-known works to its credit. Parsons Brinckerhoff has designed hundreds of projects including tunnels, dams, super highways, transit systems, and bridges. Among the best known to the public are: New York City's first subway; the Garden State Parkway; Florida's Sunshine Skyway; Hampton Roads Bridge-Tunnel in Virginia; NORAD Underground Defense Center, Colorado Springs; the Bay Area Rapid Transit System and Trans-Bay Tube in San Francisco; the Newport Bridge, Rhode Island; Fort McHenry Tunnel, Baltimore; and the Metropolitan Atlanta Rapid Transit System. They also served as engineers for the 1939 New York World's Fair. Thus, Parsons Brinckerhoff has left an indelible imprint on the American landscape.

This book is not, however, simply a chronicle of completed projects, but rather a history of a very successful partnership. Thus, we see that the design of large-scale engineering works is an intensely human activity involving the personalities of engineers in the firm as well as those of their clients. The firm has seen a succession of gifted engineers/partners who have piloted their partnership through the shoals and reefs associated with two world wars and the Great Depression to remain, a century after its beginning, one of the nation's leading engineering consulting firms.

We live in a technological society which has transformed the nation into a powerful urban industrial society. This transformation was made possible by scores of engineering firms—including Parsons Brinckerhoff—who were and are engaged in the design of large-scale engineering works of great complexity. These engineers are among the real revolutionaries of our time, though they have gone about their business with little acclaim, critique of their work, or even understanding on the part of the public. There has been a paucity of histories of the practice of engineering in America since the Civil War. It is hoped that this fascinating history of Parsons Brinckerhoff will provide new perspectives on how engineers go about their business.

EMORY L. KEMP, Ph.D
Director, Program for History of Science and Technology,
West Virginia University

AUTHOR'S PREFACE AND ACKNOWLEDGMENTS

Sometime in 1931, when the firm of Parsons Brinckerhoff occupied cramped and relatively inhospitable quarters in New York's Maiden Lane, the son of one of the resident engineers happened upon William Barclay Parsons, Jr. on the stairs. "He was a rather tall, spare man," the boy later remembered, "with an impressive beard, who seemed to me what God must look like."

Notions of the deity aside, Parsons inspired something just short of idolatry in many men, for he had conspicuously noble qualities which set him apart: his broad intelligence and philosophic view, his engineering genius, a talent for public service, a rigorous integrity, his scholarship and literary gift. He was a man to be reckoned with on all counts, and it is not too much to say that the firm he founded — which is now in its hundredth year — has had to reckon with him ever since.

Parsons founded the firm in 1885 in New York City as a sole proprietorship; today, it is a corporation, or family of corporations, of international scope with well over one thousand employees. As such, it is one of the oldest and most distinguished companies of its kind in the United States.

This book, commissioned in January 1983, is its centennial history.

Though I had written of Parsons himself in an earlier book (*Labyrinths of Iron: A History of the World's Subways*), I have tried not to repeat myself, and where possible to organize my account of his career around supplementary material which has come to hand. In other respects, this history is based almost entirely on information in the archives of the firm.

That information, though uneven, proved substantial. I suppose at one time or another the whole PB Communications staff helped to gather it in, but those whose industry was most apparent to me are Rena Frankle, Carol Meisel, Rose Reichman, and Rebecca Yamin. With encouragement from Henry Michel, they conceived and developed the project, catalogued the documents, conducted a series of valuable oral history interviews with notable employees, and extended to me every assistance I required. The interview archive is particularly a monument to the labors of Carol Meisel, while Rebecca Yamin, through her tireless research and extensive editorial contributions, made sure no item of value or interest escaped my eye. Rebecca Yamin also prepared the four folios of photographs. Rose Reichman, with Rena Frankle's able assistance, expertly coordinated it all. To their patient effort and support I owe whatever is solid in the foundations of this book.

I would also like to thank the many partners and principals of the firm with whom I spoke for being generous with their time, full, forthright, and often quite candid in their views. I salute them for it.

Drafts of the manuscript were usefully scrutinized by the firm's Centennial Review Committee (William H. Bruce, John E. Everson, Seymour S. Greenfield, and Winfield O. Salter); by Kevin Curran, Walter S. Douglas, William T. Dyckman, Thomas R. Kuesel and Henry L. Michel; and from outside the firm by Robert M. Vogel, Curator, Division of Mechanical and Civil Engineering at the Smithsonian Institution; Benedict Leerburger, Editorial Consultant; and Alfred D. Chandler, Jr., Strauss Professor of Business History, Harvard University.

To my wife Danielle I owe a tender debt of gratitude for her sound and sensible editorial advice, and to John W. Hawkins and Alice Martell most bountiful appreciation for preparing the way.

SPECIAL ACKNOWLEDGMENTS BY THE FIRM

There are many people whose names and specific labors go unmentioned in this book but whose contributions are attested by Parsons Brinckerhoff projects all over the world. It is on the foundation of their work, in addition to the work of the partners and principals, that the firm's reputation rests. There is also work behind the scenes—on the drafting boards, behind the typewriters, and at computer terminals—that contributes to the product. All of it is essential, all of it is reflected in the firm's one-hundred-year-long history.

CONTENTS

CONTENTS

PARSONS BRINCKERHOFF
THE FIRST
100 YEARS

UPON A SURE FOUNDATION:
1885–1932

CHAPTER 1

He stood on a dock in Hankow, a stranger in a strange land, and watched with a feeling of desolation as the lights of the steamer carrying his wife and child disappeared in the darkness down the Yangtze. He had every reason to believe at that moment he would never see them again—not for any danger to themselves, for they were homeward bound, but because he had remained behind on a mission he was told he would not survive.

It has been said by more than one historian that the third quarter of the nineteenth century marked the end of the age of heroic engineering—the end, that is, not of works of spectacular magnitude or difficulty, but of "the drama of a single individual striving against odds." But this was 1898; the man was an engineer, not a saboteur or a spy; and if anything gave him courage and resolve that night it was his belief that, in the new age no less than in the old, an engineer could still be a "bold adventurer and universal man" as fit to "shape the lives of men and nations" as generals, industrial magnates, and officials of state.

Such an epic view of the calling, of course, reveals as much about the man as the profession he espoused. But William Barclay Parsons, Jr. was, as a friend remarked, "most exceptional. To him, engineering was an instrumentality of civilization, not a mere physical arrangement of physical things." Parsons himself declared: "Of all human activities, engineering is the one that enters most into our lives"; and while he approved of Thomas Telford's famous definition of it as "the art of directing great sources of power in nature for the use and convenience of mankind," he added, wisely: "There must be a background of culture to produce the mental poise necessary to accomplish such a tremendous task."

As Parsons was profoundly aware, the word "engineer" meant originally "man of character" or "man of genius." When first used to designate a profession, however, in the fourteenth century, it meant simply "one who contrives, designs, or invents" and colloquially, "a layer of snares."* There are all kinds of engineers, of course, just as there are all kinds of men. Parsons, however, was one man who deliberately cultivated his life in the profession according to the ancient ideal.

William Barclay Parsons, Jr. was born in New York City on April 15, 1859, the great-grandson of Henry Barclay, second rector of Trinity Church, and the great-great-grandson of a British naval officer whose ship had been wrecked off the Long Island coast during the War of 1812. A more distant relation was Colonel Thomas Barclay, a prominent Tory during the Revolutionary War. By pedigree he was therefore linked to the Anglophile aristocracy of old New York, a fact he self-consciously savored by adopting Barclay as his given name.

At the age of 11 he was sent to school in Torquay, England, and thereafter studied for four years under private tutors while traveling through Germany, Italy, and France. Though a continental tour was a tradition of European education, the itinerant character of these years probably owed something to the nature of his father's responsibilities as a partner in the chemical import firm of Parsons & Pettit.

The family prospered. In 1875 they moved from a modest apartment on Bleecker Street to 500 Fifth Avenue, a handsome brownstone complete with its own ballroom, and Parsons entered Columbia College.

Though in his later years he was thought, on occasion, to display a sort of humorless rectitude (as a senior classmates dubbed him the "Reverend Parsons"), his college days preserve some evidence of normal adolescent mischief. On October 2, 1876, he was called before the faculty "for making a disturbance on the college grounds," and a year later was threatened with expulsion for the "first infringement of the rule prohibiting conversation during chapel services." Otherwise, his record appears blameless. He was a founder and the first sports editor of the Columbia *Spectator,* president of the College Athletic Association, and star debater of the Philolexia Society.

He was an excellent student. In 1879 he entered the Columbia School of Mines (precursor of the School of Engineering), where his

---

*"Engines" at the time were typically of a military character. As the Latin cognate (*ingenium*) also meant "wit," they were, so to speak, devices to outwit the enemy, hence the pejorative, "snares."

only weak subjects were blow-pipe analysis and stonecutting. In 1882 he graduated with the highest average attained up to that time.

Apparently in professional demand even before his graduation, he took his final exams early to accept a position with the Erie Railroad at Port Jervis, New York. In the following year, he was promoted to roadmaster of the Susquehanna Division, and on June 3, 1884 took charge of the Greenwood Lake Railway, which was in desperate need of repair. Meanwhile he had published two technical manuals, both considered standard: *Turnouts* (the arrangement — or switch — by which one track leaves another) establishing exact formulae for their determination and providing accurate tables for use in the field; and *Track: A Complete Manual of Maintenance of Way.* He sent a copy of the latter to his Columbia professor, F. R. Hutton, who affectionately replied: "There is something very gratifying to a teaching engineer to have a graduate whose nose he helped hold to the stone come out as a leader in his specialty as you have done, and I use your success to point a variety of morals and adorn several and sundry tales to our young men of today as to what may come about for brains and diligence in the use of opportunities." At 25, Parsons was being held up as an example, and in one way or another this would be true for the rest of his life.

In May 1884 he married Anna DeWitt Reed and decided to establish an office of his own. On January 1, 1885, at 22 William Street in New York, he hung out his shingle as consulting engineer.

His brother, Harry de Berkeley Parsons, soon joined him. Their talents were complementary, but they apparently regarded each other with reserve. Harry, three years younger and a mechanical engineer, was their mother's favorite; Barclay, their father's pride. Harry was quiet, fussy, pedantic, set all his grandfather clocks to chime in unison on the hour, and confined himself more narrowly to his field. Barclay made no secret of his grand ambitions and whereas he would eventually write with authority on many subjects, Harry's one book was on *Steam Boilers: Their Theory and Design.* They went yachting together on weekends, but professionally they eventually parted ways.

Nevertheless, for 15 years they collaborated on a number of notable projects, including the Fort Worth and Rio Grande Railroad in Texas, a bridge over the Brazos River, the first railroad in Jamaica, two major water supply systems (for Vicksburg and Natchez), and several hydroelectric plants, especially the Spier Falls Powerhouse and Dam on the Hudson. Independently, Harry built a reptile house for the Bronx Zoo and supervised foundation work for the Episcopal Cathedral Church of St. John the Divine.

Such activity was gratifying enough, and certainly contributed to Parsons' reputation; but it was not the reason he had opened an office of his own. That reason, as he once told a journalist, was to make himself available for what he had already decided was to be his life's work: the building of a subway for New York.

There were plenty of potential subway engineers around, and it is striking that his ambition proved so prophetic. Over the course of at least three decades numerous subway schemes had come and gone, and one novel miniature system had in fact been built—secretly, in 1868, by Alfred Ely Beach, a publisher and inventor. This had been operated by pneumatic power in a block-long tunnel under lower Broadway, where a giant fan had propelled a little cylindrical car back and forth along the tracks. A sensation at the time, Beach's experiment was alternately checked by political intrigue and by a crisis in international finance. Instead, the city had gone on to erect elevated railroads, which by 1885 ran the length of Manhattan Island along four of the city's main thoroughfares. But like the horse-car railways and omnibus lines before them, the els were soon overcrowded—"double the capacity of cars," according to the *Evening Post,* though the trains were running "as frequently as possible consistent with safety."

London had a subway—the world's first, opened in 1863—but it was operated by coke-burning locomotives, and discouraging stories filtered back across the Atlantic of suffocations in the tunnels and of dreadful ailments induced by inhaling the sulphuric acid gas.

Nevertheless, agitation for a subway, as often as it was suppressed, sprang up anew like dragon's teeth.

In 1885 the scheme that enjoyed the most adherents was the Arcade Railway, a four-track underground system flanked by sidewalks for an arcade promenade.

Parsons joined the Arcade faction, then broke away with others to form a rival company, the New York District Railway. The Arcade's provisions for ventilation in particular had seemed inadequate. "It is not enough," said Parsons, "to make a hole in the roof."

The District promised a sounder design. It did away with the arcade promenade, suggested a shallower excavation, and under Parsons' direction developed a ventilation plan on the "principle of the syringe" (or piston action of the trains) after the fashion of the new London tubes. His other assignment had been to map the route.

The Arcade and District sought legitimacy through the courts where their enmity led to their demise.

This was in 1886. On October 21, 1887, Mayor Abram S. Hewitt of New York in a significant public pronouncement singled out Parsons as the man to consult about subways; and in 1891, when a Rapid Transit Commission was at last convened to adopt subway plans, Parsons was made deputy chief engineer.

The Commission failed miserably: not a single bid for the franchise was received. However, by 1894 some action appeared inevitable and when the Commission reformed it chose Parsons as chief engineer. "I was 35 years old at the time," he later recalled, "and when I look back now I am glad I was not older. I doubt if I could now undertake or would undertake such a work under similar conditions. But I had the enthusiasm of youth and inexperience. If I had fully realized what was ahead of me, I do not think I would have attempted the work. As it was, I was treated as a visionary. Some of my friends spoke pityingly of my wasting my time on what they considered a dream."

Parsons' youth and inexperience, in fact, were held against him, and to strengthen his credentials he immediately set out on a personal inspection of rapid transit abroad. He traveled to England, France, Germany, and a half dozen other countries where subways were being planned, and in his comprehensive report weighed numerous controversial questions in the scales. Among other things, he decisively argued for electric traction over steam.

The superiority of electricity was by no means yet obvious, and opposition to steam was customarily advanced more on environmental than technical grounds. However, as Parsons painstakingly showed, a rotary electric motor taking its power from a distant station is ultimately more efficient than a reciprocating engine which wastefully carries and consumes its own fuel.

Largely on the basis of his report, the Commission of 1894 established the major guidelines for the subway that would be built in New York.

The Commission had the power either to sell a franchise to a private corporation (as the 1891 Commission had hoped to do) or to involve the city itself in rapid transit construction and ownership. Parsons opposed the second alternative, because he was "against all socialistic tendencies,"* though in the November election public ownership was favored by three to one. Nevertheless, his fears were soon allayed, since the terms of the franchise made municipal ownership in practice rather remote:

> the city would pay the contractor the amount of his bid out of funds provided by city bonds; he in turn would pay as rental the interest on the bonds, plus one percent per annum as a sinking fund sufficient for their payment at or before maturity. At his own expense, he would also provide rolling stock, powerhouses, and other equipment—the infrastructure—so as to indemnify the city against loss, but in turn enjoy income from the lease which would last for fifty years.

Though at its conclusion the city could purchase his investment, he could also renew the lease if he wished. Parsons, who thought the subway would run at a profit, assumed he would.

This contract formula, however, depended on the loan of the city's credit—and here an obstacle arose.

> The extent of credit rested on the city's debit limit, as determined by law, and in 1894 it was clear that the credit needed to match the probable cost

---

*But not against all Socialists. Many years later, on a holiday in Scotland, he was out walking with two of his grandchildren on the moors when they passed a dignified old gentleman who looked to be his twin—"same height, same salt-and-pepper beard, same cap, plus fours, tassel socks, brogue shoes." As the two men nodded to one another, the children began to giggle. After they had gone on a ways, Parsons stopped and smartly banged their heads together—"the only time," recalled one, "he ever punished us in his life." "I want you always to remember," he told them, "that you have just seen a very great man, George Bernard Shaw."

In his own way, in fact, Parsons was a bit of a Utopian. Once noting that "the laws of attraction and repulsion govern the way individuals respond to one another," he went on to conjecture: "Should the application of natural laws to human affairs ever be grasped and obeyed, then individuals, groups, and nations would pursue their several paths with respect to those pursued by others, as do the great members of the stellar universe, or the infinitely small components of the atom in their orbits, without interference or fear of collision." The beguiling symmetry of this notion is characteristic of Parsons' philosophical speculations, which were usually inspired by the aesthetic character of Nature's paradigms.

of subway construction far exceeded the amount the city could command. This one hitch, which had to be overcome before private capital could be enticed into the undertaking, held up the process for years.

Meanwhile, civic pride was embarrassed as, one by one, Glasgow, Budapest, and Boston opened subways, Paris appeared about to, and London's newly electrified system had ramified into a sophisticated network that served the city as a whole.

CHAPTER 2

Other events, however, were about to overtake Parsons' career; and they had nothing to do with a subway in New York. In February 1898 a Spanish mine blew up the U.S. battleship *Maine,* in Havana Harbor, and America and Spain went to war. Parsons promptly volunteered to raise an engineers' regiment and took charge of a training camp in Peekskill where he drilled the New York recruits. He must have drilled them rather well. Six months later, at the end of the war, he emerged as a brigadier general in the National Guard.

The Spanish-American War was a short-lived and decidedly one-sided conflict, but it marked a turning point in the nation's history— "the birth," as Parsons put it, "of the United States as a colonizing power." For aside from securing the independence of Cuba from Spain, the United States almost casually acquired control of the Caribbean through Puerto Rico, secured a naval base in the Far East through its occupation of the Philippines, and completed its hold on Hawaii. Not incidentally, the war also made construction of the Panama Canal an eventual certainty.

In a parallel development, American capital had begun to put down roots in China and Japan. Backed by King Leopold of Belgium, J. P. Morgan, and two U. S. Senators, an American syndicate had obtained a concession from the Chinese government for a thousand miles of railway from Hankow, the "Chicago of China," to Canton, her principal seaport in the south. In April 1898 the Chinese Minister in Washington, Wu Ting-fang, signed the contract, and the syndicate headed for Peekskill to persuade Parsons to survey the route.

He didn't need too much persuading. He was in a restless—possibly reckless—state of mind. The subway plans on which he had labored for more than a decade remained at the mercy of Albany intrigue, which dragged interminably on; and he was doubtless growing weary of the pitying commiserations of friends. He accepted the assignment with alacrity—even though, as he told his friend Nicholas Murray Butler, he was "confidently informed he would be killed."

China was in turmoil, its inhabitants fiercely xenophobic, and the route between Hankow and Canton crossed the "closed province" of Hunan, of the eighteen provinces then constituting the Chinese Empire the only one not yet mapped or explored by a foreigner.

But Parsons was inspired by a kind of epic dream, peculiarly American perhaps, which had recently moved him to eloquence on the spanning of the American continent by railway with the junction of the Union Pacific and Central Pacific lines at Promontory Point in Utah:

> Although in the building of these lines there was not a single piece of spectacular or outstanding designing, and nothing worth describing in textbooks of engineering except the discovery of the most favorable pass crossing the Rocky Mountains, nevertheless, in their entirety they stand unsurpassed in the art of construction. Here was a railway projected before the Civil War began and completed soon after it was ended, extending westward from a sparse agricultural settlement on the Mississippi River to the mining camps on the Pacific Coast. The intervening 1500 miles were almost a *terra incognita* . . . . To do the planning, to gather the materials, and to carry such an enterprise to successful completion demanded an extraordinary combination of vision, executive ability, technical skill, unflagging enthusiasm, and indomitable courage, and the result was not merely the laying out of a railway but the making of an empire. What calling can point to a single act of comparable achievement?

It was some such idea—the crossing of a *terra incognita*—not merely the laying out of a railway but the consolidation of an empire— which now led Parsons on. Beyond that, in his own mind at least, China and America were uniquely connected by destiny. "The world's progress," he wrote, "has always been from the rising to the setting sun *ex oriente lux*. Now, after a lapse of five thousand years, the youngest of the great nations is preparing to pass on, or rather to return, this light to the oldest, whence it started on its 'circum-orbem' journey."

It is no exaggeration to say that Parsons saw his mission as belonging to a cycle in the world's history.

So in the summer of 1898, accompanied by his wife, daughter, and five assistant engineers, he set sail for Shanghai. But even while he traversed the high seas, the crisis in China grew worse. The Dowager Empress toppled the Reformist regime and proceeded to behead its lieutenants with despotic speed. She appointed a new governor of Hunan, Yu-Lien-san, who was determined to thwart Parsons at every turn, and a new "Director Several of Railways," the most corrupt official in her coterie.

When Parsons disembarked at Hankow (after a provisioning stop at Shanghai), he was told that he could not be promised safe-conduct. His party was assigned a company of soldiers, mainly to indicate its official character, but the hundred attendants he had been promised were refused. When he let it be known, however, that if he had to he would start the journey alone, the government relented and the crew was suddenly produced.

Parsons' course for some distance from Hankow lay toward the Nanking Mountains along the Yangtze and its tributary the Siang. To avoid whenever possible the dangers of sleeping on shore, he hired a junk run by a smuggler to serve as his headquarters. The boat, wrote Parsons romantically, was one of unusual height and fit for "a gay freebooter ploughing the Spanish Main."

He progressed about ten miles a day despite touchy negotiations with local officials, supply problems, and numerous menacing delays. On one occasion when he strayed from the ranks, he was pelted with earth and stones. Though singled out as a "foreign devil" (*yang-kwei-tze*), he overcame the stigma, "pushing ahead with indomitable will, making friends and admirers of men who were prepared to distrust and hate him." For one stretch of 500 miles, he later recalled, "I was the first foreigner ever seen."

By Christmas the party had reached Ping-shui, where their Yuletide festivities struck the local inhabitants with the same sort of horror and strangeness that Westerners had often expressed toward aboriginal rites:

Our actions, our songs, our very food, but above all, our forks and knives, were a source of inexplicable astonishment to the people; but when our plum pudding—a thoughtful gift of an English lady in Hankow—appeared, decorated with holly and blazing in true Yuletide style, a look of terror appeared on their faces. The climax, however, was reached when a flash-light picture of the scene was taken. When the magnesium powder flared up, the crowd broke and ran. Probably the natives of Ping-shui stoutly maintain today that "foreign devils" are huge men with beards, who feed on uncooked meat which they tear to pieces with short swords and spears, and which excites them to such a degree that they shout loud and often, and in the midst of their excitement eat flames.

Oddness is even, however, and to a Westerner Chinese customs could not have appeared more opposite if they had belonged to the legendary people at the antipodes: in China, white was the color of mourning, left the place of honor; men wore skirts, women breeches; everyone was addressed with the family name first, read and wrote from right to left, spoke of the magnetic needle as pointing to the south ("singular for a people living in the northern hemisphere"), and even arranged the teeth on the blades of their saws to cut on the upstroke rather than the down.

Nevertheless, in the communality of peoples, fellowship finds a way. And so it was that Parsons met a local magistrate who more than shared his predilection for Old Glenlivet Scotch.

At the time of passing through his jurisdiction our headquarters were afloat, so that he joined us with his junk, and every night his place at dinner was regularly set, and on returning to his own boat he always took with him a comforting and comfortable glow. One night as he was leaving, dressed as usual in his long embroidered official robes, with his button and his peacock feather, "chin-chinning" or bowing his farewell as he walked backwards down the narrow plank connecting the junk with the shore, there was suddenly a series of rapid gyrations, like the rotating of the sails of a windmill, then a void in the night air, followed a moment later by a loud splash . . . The next evening he called as usual at the dinner-hour, and expressed his deep mortification at the previous evening's catastrophe, explaining at length that his servant . . . had held the light in the wrong place. We begged him not to mention it; that we understood the phenomenon perfectly; that our servants had been known to hold double lights, bringing us to grief, and, in fact, it was well authenticated that in our large cities, where lights were firmly fixed on iron poles, the latter have been seen to wave.

The conviviality of this official was exceptional, of course; on the other hand, hostility was not the only problem Parsons' survey faced. Local maps, even those purporting to give details, were "caricatures, and outdid the productions of Herodotus and the early European

geographers." One, for example, which depicted the familiar landscape around the viceregal capital of Wŭ-chang and the city of
Hankow, ought to have shown the Yangtze River running straight.
"Had the local cartographer so shown it on his map, one of two things
would have happened: either he would have been obliged to use a
larger sheet of paper or the river would have run off the border. He
very successfully and ingeniously avoided both difficulties by giving
the river a graceful bend."

Nevertheless, the party managed to establish the correct longitude
and latitude of various towns and cities, and among other things
discovered the true pass across the range connecting the headwaters
of the Wu-tan with the Wu-shui. In a patriotic flourish, the staff
named it "Parsons' Gap."

It would not be fair to say that Parsons' attitude was condescending,
or that he was blinded by an ethnocentric bias in both eyes. In one
eye, perhaps, on occasion—but not in both. He had prepared himself
with a study of Chinese history, philosophy, and the classics, regarded
talk of the "Yellow Peril" as perverse, and took pains to remind his
compatriots that however backward the people might presently seem,
their heritage was second to none:

> A nation that has had an organization for five thousand years; that has used
> printing for over eight centuries; that has produced the works of art that
> China has produced; that possesses a literature antedating that of Rome or
> Athens; whose people maintain shrines along the highways in which, follow-
> ing the precepts of the classics to respect the written page, they are wont to
> pick up and burn printed papers rather than have them trampled under foot;
> and which, to indicate a modern instance, was able to furnish me with a
> letter of credit on local banks in unexplored Hunan, can hardly be denied
> the right to call itself civilized.

(The bank note, in fact, had originated in China eight centuries before
its reinvention in Europe.)

Some things, however, took even Parsons by surprise: in particular,
that the Chinese "were acquainted with good engineering design." He
saw bridges "that would have done credit to any architectural engineer brought up in the most fastidious school of Europe," and wondered
if the Chinese, rather than the Romans, had not invented the arch.

More startling, he discovered that the skyscraper method of putting up buildings by encasing a rigid frame in thin masonry walls ("supposed to be something essentially American"), was common throughout the land.

The humanity of Parsons' interest made itself felt. As distrust waned, curiosity and affection grew. Others soon joined the caravan, which towards the end of its trek included 600 people proceeding in a single file over several miles. "We had shown no fear," he recalled "and consequently the people feared us; we neither molested nor interfered with anyone, therefore the people respected us; and we paid regular prices for our provisions, and would not allow our attendants to steal, therefore the people liked us."

Parsons triumphantly entered Canton two days ahead of schedule, "by a singular coincidence at noon," as he put it, on the very day he had named some months before in a letter to his wife. Underweight, his hair and beard gone gray, he was "only too glad it was over" without having lost a single man.

§

Parsons' vicarious romance with the Chinese Empire was not untouched by innocence. Despite certain misgivings, he never really doubted that the attraction of foreign capital to China was a good thing. Mining and other privileges accompanied the railway concession, and "it took small flights of fancy," he wrote,

> to see future trains bearing their dark burden northward to furnish power for the furnaces and mills that will be built in central China to convert her ores into metals or work her raw produce of cotton and wool and hemp into articles of commerce; or other trains south-bound carrying a like burden to Canton and Hong Kong to make steam for vessels of all nations, bringing goods from other lands to China's, and taking back her teas and silks.

But there were two sides to the story, and these fancies unquestionably took flight from the brighter side. Though Parsons was not a colonialist, and was bitterly opposed to wars of annexation, he was also convinced that China needed "a little help . . . to work out her own salvation." Unfortunately, colonial interests had been proceeding under the guise of "a little help" for quite some time, and from so many quarters that by 1898 the country was in serious danger of partition. And railways were drawing the lines. As one standard source sums it up:

> Russia acquired a monopoly on railway building in Manchuria; Belgian financiers . . . provided the money for the road from Peking to Hankow; Germans furnished the capital for lines in Shantung and for the northern section of the road from Tientsin to Pu-k'ov [on the Yangtze, opposite Nanking], Britons for the southern half of the road and the line from Shanghai to Nanking, and France was granted concessions for railways in its sphere of interest—Kwangtung, Kwangsi, and Yunnan. . . The French also got from the Russo-Chinese bank the contract for a road connecting T'ai-yuan with the Peking-Hankow line, and a Franco-Belgian syndicate the contract for a line from K'ai-fang to Lo-yang. . . A U.S. syndicate [Parsons' mandate] was given the concession for a road from Hankow to Canton.

When connected, these lines would quarter the empire, not, as Parsons optimistically wrote home to a friend, "actually and metaphorically bind them together with bonds of steel." Though not oblivious to schemes of division, he yet hoped for the "coming of a Peter the Great to elevate his people with the developing of industry, and a Washington to instill in them a lofty sense of national unity, spirit of freedom, and love of country." That, to his credit, was the dream he dreamed. But it takes no flight of fancy to guess how inconvenient to certain interests the emergence of such men would have been.*

---

*Parsons is remembered in China as he would have wished. Recently, an engineer from the firm was hailed as the representative of "the great Parsons" when he visited Tianjin (Tientsin in Parsons' day), and Parsons' book on China (long out of print in the United States) has been reprinted in Taiwan.

CHAPTER 3

Had Parsons perished at the age of 40 beneath a hail of stones in Hunan, he would have died a man of whom great things were expected, yet with nothing of lasting importance attached to his name. In his own view, in fact, he would have died a failure, "foolishly obstinate," as he feared, in pursuit of his great life's work which lay unrealized still among a folio of parchment tracings buried in a dusty drawer.

Parsons' China adventure marked the beginning of his true renown. And it seemed to change his luck. No sooner had he arrived in Canton than he received an urgent cable from New York: return at once, he was told, the subway is at hand.

In his absence Abram S. Hewitt, the architect of the overall contract formula, had cut the Gordian knot: the debt limit of the city had been raised by a constitutional amendment, and the valuation of its taxable properties against which it could borrow as security had been increased.

The subway contract was awarded to John B. McDonald, who had proven himself in a number of large undertakings, for a bid of $35 million. McDonald, however, failed to come up with the security, and had to relinquish his prize to August Belmont, Jr., the financier and American representative of the Rothschilds. Belmont retained McDonald as general contractor, and organized two companies, one within the other, to take charge of the contract's different obligations: a construction company to furnish the security, and the Interborough Rapid Transit Company, or IRT, for the lease and operation of the road.

Belmont and Parsons were friends. They had collaborated briefly on the New York District Railway 15 years before, and Belmont had invested (cautiously) in the China venture. And they would work together again.

Belmont was a major power in the land. In addition to his own banking house, he directed the Equitable Life Assurance Society, the Bank of the State of New York, the Manhattan Trust Company, the

17

National Park Bank, and the Guaranty Trust Company of New York. He was obviously a financier who could afford to take risks; even so by his own account, it was his personal confidence in Parsons that clinched his commitment to the IRT.

Parsons designed the route in the shape of a "Z." It began with a loop near City Hall, proceeded north with four tracks up to Grand Central Station, then westward to Broadway (now the shuttle line), and along Broadway to 104th Street. From there one double-track line continued north along Broadway to Kingsbridge, and another eastward under the Harlem River through the Bronx to the Zoological Gardens. In all, its length would be about 21 miles, including five miles of elevated track, to be built in four and a half years.

Ground was ceremonially broken on March 24, 1900, in front of City Hall. Thousands milled about as fireworks burst overhead in brilliant geometric displays, and John Philip Sousa struck up his famous band.

Mayor Robert A. Van Wyck dug the first shovelful of earth with a Tiffany silver spade, and in keeping with his knack for playing the buffoon, "deposited the load in his silk top hat." As he straightened up "with a benignant smile," cannons thundered, church bells chimed, and boats and ships in the harbor whistled and wailed.

In his speech Van Wyck compared the subway in importance to the Erie Canal. Privately, however, he regarded it as a personal defeat. When he had come into office as the Tammany Hall candidate in 1898, he had tried to destroy the Rapid Transit Commission and requisition its plans. Parsons, anticipating this, had told the counsel of the Commission before leaving for China: "Try to arrange if possible that the drawings be forgotten."

But with the blueprints now spread out before him, he bent to the work at hand.

The subway, he conceded, was not to be a feat of conspicuous glory "like the construction of a magnificent bridge . . . but the securing of a simple construction from a mass of most intricate detail."

The method was cut-and-cover, which meant opening the street to the level of the utilities (about 14 feet down), as opposed to boring a tunnel beneath them by shield. The utilities were removed or otherwise rearranged, the sidewalls braced by piles, beams framed to the piles for the roof, the "cut" planked over for traffic, and the rest built underneath. Once the structure was complete—with its ribs of steel, its walls, roof and floor of concrete, and rows of steel columns erected along the tracks as intermediate support for the roof—the planks were removed and the surface of the street restored. Roughly speaking, the subway was a rectangular box built in a trench.

As a covered way, its great convenience was easy access by stairs, while gloom in the stations could be somewhat dispelled by natural light through grates and deadlights above.

There were, in fact, only three miles of true tunnel on the line: along Park Avenue between 34th and 42nd Street, where the subway had to pass below a tunnel already in use; under the northwest corner of Central Park; in Washington Heights; and under the Harlem River.

Beyond 120th Street on the West Side, the subway became a viaduct spanning the Manhattanville Valley—once known as the Hollow Way.

Cut-and-cover was not new, though Parsons has sometimes been credited with its devising. The real pioneers were Sir John Fowler and Sir Benjamin Baker, who had used it for the first London underground lines. Parsons owed much to them both, as he was the first to admit. And what he owed them was this:

> what precautions are necessary to ensure the safety of valuable buildings near to the excavations; how to timber the cuttings securely and keep them clear of water without drawing the sand down from under the foundations of adjoining houses; how to underpin walls and, if necessary, carry the railway under the houses and within a few inches of the kitchen floors without pulling down anything; how to drive tunnels, divert sewers over or under the railways, keep up the numerous gas and water mains, and maintain the road traffic when the railway is being carried underneath; and finally, how to construct the covered way so that buildings of any height and weight may be erected over the railway without risk of subsequent injury from settlement or vibration.

It was Parsons' refinement and adaptation of the method, however, that became the standard.

That standard meant ingenuity every step of the way. As Sir William White of the Royal Commission on London Traffic observed with admiration at the time: "There was an almost infinite variety in the ways with which local conditions were met."

In construction, however, cut-and-cover work wreaks havoc on urban life, and "stretches of the route looked like the aftermath of a bombing."

But there were unexpected rewards: in the course of their efforts, workers came upon subterranean springs, brooks, and ponds and up in Harlem, a long-forgotten lake. Mastodon bones were unearthed, chests of coins, and—near the Battery in "made ground"—the charred hull of a Dutch merchant ship which had caught fire and sunk in 1613.

Inevitably, with all the blasting required, there were accidents; at least fifty men lost their lives, with many more injured or maimed.

In all, about 3.5 million cubic yards of earth and rock were removed.

Parsons brought the work in almost on time and within cost. "In the organization of his staff," wrote the *New York Times,* "personal considerations were dismissed and 'influence,' social or political, had no weight."

The subway opened on October 27, 1904. In a remarkably terse and matter-of-fact announcement, Parsons said: "Mr. Mayor and Mr. Orr,* I have the honor and very great pleasure to report that the Rapid Transit Railroad is completed for operation from the City Hall Station to the station of 145th Street, on the West Side line."

The large audience broke into applause, waving handkerchiefs and cheering. According to a journalist, Parsons "blushed like a schoolboy as the cheering lasted much longer than the speech."

The first day, seventy thousand people took the plunge. In their general excitement there were melees, and in Washington Heights police reserves were called in and formed a flying wedge to drive the people back.

Though the subway could traverse the island in thirty minutes, not everyone felt it had improved their lives—especially when advertisements began to appear on station walls. But whatever was taken to be amiss was not laid at Parsons' feet. In a typical encomium, the *New York Times* declared: "New York City will ever hold Mr. Parsons in high respect, not alone as an engineer, but as a gentleman who has established the fact that great public works may be carried to completion with an unsullied reputation and clean hands."

§

In the scrimmage for new subway lines after 1904, Parsons played an ancillary role. The major contestants were Belmont and McDonald,

---

*Alexander E. Orr, chairman of the Rapid Transit Commission.

though as the leading expert in the field it was inevitable Parsons would eventually have to take sides.

Belmont claimed that the Rapid Transit commissioners had a "moral agreement" to extend his IRT up the east side of the city and down the west, turning the "Z" configuration into an "H." McDonald, however, organized a large construction and holding company out of the Brooklyn Rapid Transit and Metropolitan Elevated railroads and bid for an alternative Third Avenue route. For whatever reason, the commissioners leaned his way.

Now, it happened that Belmont held the franchise to the Steinway Tunnel, which, if built beneath the East River, would connect Manhattan with Queens. Moreover, if connected to his IRT, it would constitute an East Side branch line and preempt (at least in part) McDonald's clientele. Belmont warned that if McDonald prevailed he would build it. In fact, he said, he would build it anyway, and the city couldn't stop him, because "I own the bed of the East River from the foot of East 42nd Street to Long Island City in Queens."

This took considerable *chutzpah,* to say the least; but his claim was upheld in court.

McDonald took another tack. He said the Steinway Tunnel would literally undermine his line because at one point it would dip below it, rendering both unsafe. Belmont rejected that as nonsense—and Parsons came to his defense.

Both Belmont and McDonald went forward with their plans, only Belmont was thought to bluff. The reason was simple: in 18 months the Steinway Tunnel franchise would run out, and almost no one believed the tunnel could be built in that time.

Belmont appealed to Parsons, who took up the dare. He erected a large working platform in midstream on Man-O-War Rock, sunk two shafts to create four new headings in addition to those on the opposite shores, increased the size of the reef with earth dug up from the shafts, installed power machinery on the landfill (for the compressed air), and drove from six separate headings at once.

McDonald saw he had lost, and let Belmont buy him out. In turn, the commissioners relented and conceded Belmont his "H."

The Steinway Tunnel (now known as the Queensboro) was completed in the summer of 1907, and on September 23, the first train sped through.

CHAPTER 4

Meanwhile, in late 1904, Parsons had resigned from the Rapid Transit Commission and moved out of its headquarters at 820 Broadway. He rented a small space for himself on Pine Street with a handsome colonially appointed office on the first floor and a little attic above for a drafting room. The attic was reached by a steep ladderlike staircase that went up through a hole in the ceiling. His only employees were two Norwegian immigrants, Soren Thoresen and Otto Andresen, and a story (told around the firm for years) has it that after Parsons signed his first major contract and was showing the client to the door, he rather grandly called up to Thoresen: "Put the whole staff on this at once."

However, common sense would suggest that by 1905 it would have been quite unnecessary for Parsons to try to impress a client in this way. One measure of his stature at this time is that he was a consensus candidate for the presidency of the American Society of Civil Engineers. Alfred Noble, a colleague with whom, interestingly enough, he had had a sharp exchange in 1903 over the preliminary design of the Steinway Tunnel, privately took him aside at a meeting and as much as told him the position could be his—if he would only seek it. Parsons' reply (June 12, 1905), which he stamped "Confidential" with an exclamation point, is striking:

My dear Mr. Noble:

I have been thinking over a great deal what you said to me the other night in the rooms of the Society, and I appreciate most deeply the very kind allusions that you made; however, I am not anxious for office. I realize that the office of president of the American Society is the highest obtainable honor in my profession in this country, and yet I honestly do not seek it. If it should come to me later, and quite unsought, it is of course an honor that no man should refuse; but to be entirely frank with you, I would rather not put myself in the way of securing it. This I do for two reasons: first, as I told you before, I do not care for office—I would rather be allowed to serve in the ranks. The other reason is that I have, so far as the Society is concerned, never done anything to entitle me to such consideration at its hands. I am not a regular attendant of the meetings; I have written but few papers; I never go to the conventions—in short, I am not one of those who have been zealous workers for its welfare. There are other men who have been, and other things being equal, I think it is only right that those who have worked to build up the Society should receive such rewards as the Society itself can give.

From a professional point of view I have been successful. It has been my good fortune to meet with openings, and things that I have undertaken have turned out successfully, in some measure, possibly due to my own efforts, but in other cases due to happy combinations of circumstances. There have been other men who have been less fortunate or less successful, through no fault of their own, and my ultimate election to the presidency would mean that some one of those men would be deprived of an honor that he would covet much more than I do, and which in many respects would be probably more deserved. . . .

I thank you more than I can tell you for your thought in speaking to me the other night as you did, which made a very deep impression. I have made several attempts to write this letter but have desisted each time, and I now ask you to accept it in the frank and friendly manner in which it is written, as I feel sure that you will appreciate my motives.

Again assuring you of my earnest appreciation of your thought, I am,

Very sincerely yours,

Wm. Barclay Parsons

§

Andresen didn't last long in Parsons' office. He had talent, but was apparently lazy and drank. Thoresen, on the other hand, abstained from both liquor and tobacco on principle and, according to his son, "couldn't wait to get to work."

A veritable "Viking of a man" with enormous hands, whose father had been in the Gold Rush, Thoresen had abandoned his technical training in Norway in 1896 and gone to sea for a year shoveling coal on a tramp steamer. "That year," he once told a colleague with a smile, "I fought out whatever meanness might have been in me."

Permanently chastened (his disposition was said to be sweet), he subsequently took degrees in electrical and marine engineering, and in 1903 settled in the United States. Though he quickly found work in his field, he wavered in his career. What he really wanted to be was an opera star. He had a beautiful tenor voice, sang comprimario roles at the Metropolitan Opera, and on one tour as a recitalist in the Minneapolis area gave over 30 concerts. But his prospects at best were precarious, and after marrying his accompanist, he resolved on engineering to make ends meet.

Parsons hired him as a draftsman, which was a great stroke of luck for the firm. Endowed with an almost perfect memory (he could even recall questions from childhood exams), Thoresen brought a miniaturist's deftness to bear on design, and on a scrap of paper like the back of an envelope could render an entire project to scale. Moreover, his off-the-cuff estimates of cost were generally right.

Though his knowledge of money was certainly keen, it was in heaven that he laid his treasures up. "Everything in my father's life," said his son, "was based on a tremendous belief in God." He got up every morning at dawn to read the Bible, was a devout Christian Scientist, and designed three churches for his faith. One year he gave one of them an organ that cost him a third of his salary. The IRS questioned the deduction, then profusely apologized.

In what makes almost a parable of his life and gifts, in his beautiful small handwriting he once copied the Lord's Prayer on a dime.

The next prominent addition to Parsons' staff was Eugene A. Klapp, a fellow Columbia alumnus* and engineer veteran of the Spanish-American War. Klapp had also been Parsons' colleague on the IRT as chief engineer of the fourth or northernmost division, which included all bridges and elevated sections, most notably the majestic viaduct that spanned the Hollow Way. For the latter, Klapp had devised a new and celebrated form of the truss arch, and with Parsons had been named on the bronze commemorative plaque in City Hall Station.

Briefly thereafter, Klapp had served as chief engineer of the Brooklyn Heights Railway Company.

A fastidious man, tall, with an exact moustache, Klapp was "at heart an architect," with a passion for antiques. In his early thirties he

---

*Class of 1889, though he didn't receive his degree until 1914. In his enthusiasm for varsity crew, he skipped his final exams to compete in a race.

had founded and edited a magazine called *The House Beautiful,* where under the pseudonym of Oliver Coleman he published articles on the design of windows, doors and rooms, the care and shaping of hedges, and other aspects of the "art and artisanship" of domestic settings. He particularly loved Cuban architecture, and particularly hated the mansard roof "and the whole period of American architecture to which it belonged."

Parsons obviously thought much of him, made him deputy chief engineer of the Steinway Tunnel, and took him into partnership. The firm became known as "Barclay Parsons & Klapp," and moved to 60 Wall Street. Meanwhile, Parsons had been appointed to the Royal Commission on London Traffic (where one of his colleagues was Sir Benjamin Baker), and to the Isthmian Canal Commission, which President Theodore Roosevelt had empaneled to study the prospects for a canal linking the Atlantic and Pacific Oceans—either directly across Nicaragua, or by a winding cut through the Isthmus of Panama. Subsequently, in 1906, he sat on the 13-member Board of Consulting Engineers to advise on the choice between a canal with locks or a sea-level route. The board came down on the side of the latter, and in this Parsons stood with the majority. The President, however, favored a lock canal, and by a narrow vote the U.S. Senate went along.

Parsons, however, was convinced that a canal without locks could be built between two bodies of water where considerable tidal differences existed, and was even then at work on a project that would ultimately prove him right.

Once more Parsons and Belmont joined hands—this time for the Cape Cod Canal.

CHAPTER 5

The Cape Cod Canal joins Cape Cod Bay, Massachusetts with the waters of Buzzards Bay to the south. As originally designed, it was a 13-mile waterway including approach channels with a depth of 20 and a width of 200 feet. At opposite ends Parsons had what he wanted: a significant difference in the time and height of the tides.

There were all sorts of reasons for constructing the canal, and some had been urged since the colonists first arrived.* But the least of them was what challenged Parsons most, the practicality of a sea-level route.

In the first place, since the Revolutionary War, when a British blockade had hemmed in Washington's Continental navy, the canal had been regarded as a military necessity — as again, when the blockade was repeated during the War of 1812. And then there were the commercial advantages, which had increased over the years. By the 1890s more than 30,000 ships a year rounded the Cape carrying raw materials and produce north and high-grade manufactures south. The strings of coastal barges were often so long that from the tug boats in the lead "the last ships could not even be seen." But when they reached Nantucket Sound, their orderly progress was broken and their fate given over to "a mariner's nightmare of shoals, twisting channels, surging currents and tides, ice, fog, and a long lee shore without ports of refuge." The Sound was a historic graveyard where, between 1875 and 1903, 687 ships were wrecked. Every fishing village and coastal town from Boston to New Bedford to Plymouth to Newport had sons and fathers who had perished in its depths. Nor were the stormy waters eastward off the Cape more kind. Taken together, they formed "the most dangerous coast to be found on the map of the United States."

"Personally," said Belmont, "should the Canal serve no other pur-

---

*It was probably the first "public work" proposed in the New World — by Miles Standish, ca. 1625.

26

pose than the saving of thousands of lives, I shall feel my own efforts are repaid."

Many charters for canal construction had been granted over the years, and alternative routes so often and minutely surveyed that, remarked one journalist, "every grain of sand had been made the victim of an algebraic equation." Yet (not unlike the subway in New York), competing political, commercial, and other interests had thwarted the project as effectively as a British blockade. "If ever a strip of land," wrote the historian Henry Kittredge, "was a parade ground for surveyors first and a battleground for legislative vituperation afterwards, it was the route of the Cape Cod Canal."

The beginning of the end of delay came in 1904, when DeWitt Clinton, a beer magnate, managed to interest Belmont in the project as a profit-making toll facility. Belmont in turn spoke to Parsons, who eagerly took it on.

Parsons arrived on the Cape in February 1906. He consulted previous studies, walked the alternative routes, said he saw "nothing formidable" in the undertaking, and concluded "it would not be a difficult thing."

He was wrong.

The route chosen was the logical one, a winding path northeast-southwest through a broad valley where the isthmus joins the mainland, and where Indians hundreds of years before had dragged their small boats and canoes from bay to bay. Of all the routes, this had always been the most favored, though because of the peninsula's striking configuration, the options were fairly broad. Thoreau had written: "Cape Cod is the bare and bended arm of Massachusetts; the shoulder is at Buzzards Bay; the elbow or crazy-bone is at Cape Mallebarre; the wrist at Truro; and the sandy fist at Provincetown." Anatomically speaking, Parsons cut it off at the scapula bone.

The Steinway Tunnel was just holing through when ground for the canal was broken on June 22, 1907, at Belmont's ancestral farm in Bournedale. In a ceremony oddly reminiscent of the IRT's, Belmont took a tiny silver shovel provided by Tiffany's and "dug a little dirt and moved it a few feet." He recalled the IRT in his speech: "It is both a poor and stupid argument that the past failures to build this canal should still nurse skeptics on the subject. The subways in New York went through the same way for 20 years or more. Our engineer,

William Barclay Parsons, is just as sanguine about this as he was about them, and so am I."

The comparison was more apt than he knew. It took a long time for the construction company to buy the land for its right of way, and longer still for the proper equipment to arrive. And when it did, it didn't work very well. Two land excavators, for example, with eight-ton buckets "to scoop off the overburden in the dry," turned out to be ungainly, and were dismissed by Parsons as "most unsatisfactory . . . not adapted to the soil." One costly dredge, arriving late, had to weather the winter in Provincetown.

At about the same time, Parsons discovered that the shallows near the entrance to the canal in Buzzards Bay were clotted with boulders, "huge chunks of glacial debris." These would be difficult to dredge and were bound to require blasting, so to avoid them as much as he could "he designed a channel with two turns that followed the contours of the shore—inside Mashnee Island and through Phin-neys Harbor, with its entrance off Wing's Neck." Meanwhile, at the eastern end in Cape Cod Bay, granite blocks from quarries at Blue Hill, Maine, were being dropped from scows into 40 feet of water to build up a mammoth breakwater against waves from northeast-ern gales.

The steam-shovel digging on land went smoothly enough. But a canal requires continuity of depth; and it was in dredging its path far out into open waters that the more difficult problems arose. In the winter of 1911 some dredges got stuck in the ice, and though five were subsequently withdrawn, one that remained destroyed a wild oyster bed, for which the company paid dearly in court. Moreover, despite the studied double curve in Buzzards Bay, the dipper dredges strug-gled with unexpectedly stiff clay, and the boulders Parsons thought he could avoid were found to lie in nests across his path. Some of them weighed over 100 tons, and when the dredges failed even to budge them, "divers had to be called in to shatter them with dynamite."

There were other problems. It was found that the county commissioners, who were allowed a fee for their time, were submit-ting padded bills. "If one of them lay ill in bed," complained Henry Welles Durham, the resident engineer, he still insisted on payment on the grounds that "he was lying there thinking about the canal." More seriously, it dawned on Durham that he had been shortweighted on his breakwater granite from the start:

> The first bit of rock was part of a load of one hundred twenty-five tons. . . . The contractor claimed one hundred fifty tons and, as my estimate was based on bow and stern readings in fairly rough weather, I let it stand. . . . What seemed curious was that the subsequent errors were all in the contractor's favor. . . . The difference between his bills of lading and our measurements increased until all his captains knew we were suspicious and one old Maine sailor when asked by our inspector how much he had in his latest load said, "Wa'al, the bill of lading calls for one thousand tons. If I had it aboard I'd sink."

Nevertheless, the work was getting done, slowly to be sure and stone by stone, but at last the breakwater stood: eight feet high above high tide, and 3000 feet across out into Cape Cod Bay. By early 1914 the canal was essentially cut and the channels dredged, while the company had also built a ferry slip at Bournemouth and three bridges over the canal: a single-leaf bascule for the New Haven Railroad at the mouth of Buzzards Bay; a highway drawbridge at Bourne with two 80 foot leaves; and a small draw span at Sagamore.

On April 21, 1914, with one dike-protected strip of land left, Belmont gathered the principals together for a ceremonial "blending of the waters." He took water in bottles from the canal on either side and poured it out together so it mixed while falling to the ground. He praised Parsons as "resourceful and tireless," let a little stream trickle through the dike, and then reached across it to shake hands with his chief engineer.

On July 4th the last barrier was cut. A wedge was taken out of the dike, "and the high tide from Cape Cod Bay did the rest. The water poured through, chewing a huge gash through most of the dike and churning trees, boulders, and sand. It was a wild, frightening scene, the water moving like a millrace at flood time."

The force of the flood might have given Parsons pause. In Cape Cod Bay the mean high-tide level was five feet higher than at Buzzards Bay, where the tide came earlier by more than three hours. This means that the current changed direction every six hours, and was considerably stronger going west. Though its strength and shifting force (as Parsons predicted) would scour the channel and, in winter, prevent the formation of ice, he had not expected it to be as swift as it was. It reached 7.6 feet per second, and at times actually appeared to be running uphill. In a sinuous, relatively narrow channel, which especially around Bournedale was shrouded in vapor or "ground fog," that meant danger.

Howard Elliott, president of the New Haven Railroad, would later claim he had been told the current would not be so much that "a barrel would float through;" and in demanding reparations for the "ruin of his railroad," predicted that ships would collide with the Railroad's bascule bridge and "carry it away."

The canal opened on July 29. A substantial fleet had gathered in New Bedford, while shops and homes along the Cape were decorated with flags, and thousands of people lined the shores. The *Rose Standish,* a Boston excursion steamer with Belmont, Parsons, and other dignitaries aboard, entered the canal from Buzzards Bay, followed by destroyers, yachts, and a tugboat for the press. Off Wing's Neck, two Navy submarines were standing by. At 1:31 p.m. the *Rose Standish* "knifed through a piece of colored bunting hung across the canal," and after proceeding eastward to Sandwich with the current, returned westward with the change in tide.

But as the current increased to five knots the *Rose Standish,* "with paddle wheels thrashing in reverse," was ominously "swept by a small dock at Bourne Neck and far out into Buzzards Bay."

Potentially at least, the canal was a great waterway—as the future would prove. But at the time its disappointments largely obscured its worth. It had taken almost as long as the Panama Canal to build (their dates of construction nearly coincide),* was four years late and over budget, narrow, shallow, awkward to navigate, and fit only for lesser vessels to use—and then only in one direction at a time—and then only with a favoring tide. Time and tide may wait for no man, but a great deal of time was lost in waiting for the tide to change. Accordingly, ship captains, discouraged by its inconveniences and wary of its own peculiar risks—for which, after all, they had to pay a toll—continued instead to brave the shoals and stormy seas. As a result, the traffic was

---

*The Panama Canal was built in 1904-1914. Its length shore to shore was 40.27 miles; from deep water to deep water, 50.72 miles.

a fraction of what had been predicted, while the local industry that was supposed to develop in its wake never did. Not one of the subcontractors made money, the general contractor went bankrupt, and Belmont himself eventually lost one million dollars in cash.

It was a financial disaster.
But that is not the end of the story.

CHAPTER 6

P arsons' New York office had meanwhile been thriving; and among others, there had been two important new additions to the staff. In 1906 he had hired Henry M. Brinckerhoff, a leading traction engineer from Chicago and coinventor of the standard third rail; and in 1908, Walter J. Douglas, a master of masonry bridges from the Corps of Engineers in Washington, D.C.

Brinckerhoff was a meticulous man, "a bear for detail." Of Dutch and Scottish descent, his prosperous lineage had included bankers, shipping magnates, and realtors—though in a peculiar lapse of pecuniary foresight (or some excess of company spirit) he had sold his third-rail patent to General Electric, his employer, for $12.50. Perhaps he had this incident in mind when he later remarked that knowledge of his family's progress during 300 years in the New World gave him courage in times of stress.

Brinckerhoff's third rail was first demonstrated in Chicago at the Columbian Exposition of 1893, where it was applied to an intramural railway that carried visitors on a 20-minute circuit of the grounds. Among the technological wonders and entertainments on display were "Yerkes' distance-defying telescope," the first Ferris wheel, and Edison's "talking movie kinetograph." But the keynote of the Exposition (which celebrated the 400th anniversary of Columbus' discovery of America) was electricity, as signified by the dramatic act with which it opened. Hundreds of miles away, in the White House, President Grover Cleveland pressed a button that set in motion an engine in Chicago that turned the power on. That engine, the creation of George Westinghouse, with its alternating-current generator, would become the basic tool of the electric power industry.

Brinckerhoff's invention (in collaboration with Charles Macloskie) consisted of sliding shoes mounted at intervals along the underside of the car and in contact with an electrified third rail. The rail ran parallel to and a little above the tracks, and current flowed from the shoe through copper conductors to the motor of the train.

In combination with electrification, for urban railways at least, it sealed the doom of steam.

Like Thoresen, Brinckerhoff was large and stout; and like Thoresen he was scrupulously dry. He liked to tell stories, but they were "always clean." He had a well-scrubbed face with bright, rosy cheeks, and sandy hair that turned snow-white in his middle years—"a little, in appearance, like Santa Claus." But the epithet that has remained linked to him is not "jolly" but "dignified," which he appears to have been in a very formal way. He stood up when his secretary entered the room, and stood and bowed when she left; he even insisted on pulling out for her the little stenographer's drawer.

Brinckerhoff's whole training and experience before coming to the firm had been in electric railways, and it was primarily as an expert in this field that he excelled.

In the hierarchy of the firm, Douglas struck an alternate note. Easy-going, mischievous, more broadly educated perhaps, and catholic in his interests, and a trifle irreverent, he reportedly once remarked, "Brinckerhoff doesn't really have any problems—he has all the answers, either from his church or convention." And to Parsons himself he once quipped: "Barclay, you're not running an office but a shrine."

Born in 1872, Douglas had graduated from Lehigh University, worked as a mining engineer and a draftsman, and in 1898 became assistant engineer of bridges and highways for Washington, D.C. In the nation's showcase capital, this was potentially an important position, and to tutor his judgement for the work ahead he went to Europe to study certain outstanding bridges, "the method of their building and what made them so beautiful." In the process, he also acquired proficiency in structural mechanics and the behavior of soils.

In Washington, each bridge that he built was distinguished in some way by pioneering innovations in the use of concrete. Perhaps the most famous was the Connecticut Avenue or Taft Bridge over Rock Creek Park, which spans a deep ravine on seven full center concrete arches, and at the time was by far the longest concrete bridge in the land. Others were the Francis Scott Key Bridge, and the 16th Street Piney Branch Bridge. The latter was the first to be built in the form of a parabolic arch, and consisted of two parallel concrete arches and a roadway deck supported by spandrel beams.

Each bridge was also tastefully ornamented with lions, tigers, and other sculpted figures as befitted their neoclassical design. Overall they successfully managed the difficult aesthetic transition from 19th-century iron trusswork to the heavier structures required by the traffic of the new age.

Douglas, however, came to feel that his true professional opportunities no longer lay in public employment, and one day he went out and bought a new suit and showed up at Parsons' door. The investment certainly paid off. Parsons took a liking to him immediately because, he claimed later, "it was the first time an obviously intelligent engineer came in to see me dressed like a gentleman."

(What sort of figure, one wonders, had Brinckerhoff cut?)

In any case, Parsons took Douglas on as a consultant for two bridges in Baltimore (the Edmondson and the Wyman Park), and two years later gave him a permanent position as head of structural design.

Douglas' obvious intelligence included, among other things, the ability to perform arithmetical problems in his head for which the average engineer required pencil, paper, or slide rule. This made him a canny negotiator, and an adept at bargaining on the street. In haggling with shopkeepers, he liked to narrow the spread to within a tempting sum, then risk the difference on the flip of a coin.

His most conspicuous failing was his absentmindedness—apparently a family trait. One day he came into the office wearing one brown and one yellow shoe. Said Brinckerhoff with astonishment, "What kind of shoes are you wearing?" Douglas looked down at his feet and said, "Oh, I'm sure I have another pair like this at home."

In old photographs his face looks something between impish and owlish, but it is difficult to get a picture of him physically. In an amusing discrepancy, one colleague has described him as "slightly built, modest, and soft-spoken," another as a "short, stocky man with sharp eyes and a sharp voice." Probably he could variously appear as he wished. During work on the Cape Cod Canal he went out on the dredges, spoke to the dredging captains, "and got them to do things," he confessed to his son, "that I would not have had the nerve to do myself."

Certainly he inspired great loyalty and respect. Of Parsons' opinion of him there can be no doubt, for he progressively entrusted the direction of the firm to his hands.

The direction of the firm had been toward diversity, and by 1914 its competence covered the field. While the Steinway Tunnel and the Cape Cod Canal were in progress, Parsons' small but incredibly industrious office had undertaken shop designs for the Magor Car Company; bridges for the Pittsburgh Shawmut & Northern Railroad; the Almendares Bridge in Cuba, the first concrete arch bridge in Latin America; pier designs for the Havana Docks Company; the Colliers Hydroelectric Dam and the McCalls Ferry Power Plant and Dam, both on the Susquehanna River; Rock Creek Dam in Texas; studies for a subway in Detroit; a yacht landing at Glen Cove for J. P. Morgan (with which he was displeased); the Zapata Swamp Reclamation Project in Cuba; design of the National Steel Car Manufacturing Plant in Ontario; studies for the Mohawk Hydroelectric Power Plant; the Lake Champlain & Moriah Railroad Bridge; studies for the Springfield Union Terminal; the Salmon River Hydroelectric Plant; and a pier at Puerto Principe, Haiti. This last marked the first use in the tropics of an apron made of chicken wire and concrete for protecting wooden piles.

Another project of note was the design of a water supply system for Booker T. Washington's fledgling Tuskegee Institute in Alabama. At the time, water was obtained from a series of contaminated springs located along the bottom of a gully. Parsons, aware of the Institute's financial hardship, offered his services without fee.

Beyond a general collaboration, the division of labor among the principals was roughly along the lines of expertise: Thoresen did the drafting, Klapp and Douglas the bridges, Brinckerhoff and Parsons the transportation, Parsons and Douglas the power plants and dams. Much of the hydroelectric work had grown out of an idea Parsons had that peak power could be furnished more cheaply from electric power than from steam plants. According to a colleague, he was unable to sell this idea to steam-minded utilities at the time, so he promoted and constructed a peak-power plant as a model on Garoga Creek in New York. Its success led to the construction of the larger plant on the Salmon River for utilities in northern New York State.

Water power and supply was ultimately the field in which Parsons would end his engineering career. But before he could explore it more deeply his life in the profession took another remarkable twist.

CHAPTER 7

T he very day the Cape Cod Canal had opened and the ships went sailing through on their festive parade, Austria-Hungary declared war on Serbia. Within a week the "lights were going out all over Europe," and the entire continent was engulfed in the First World War.

In the spring of 1917 the United States entered the conflict, and "as a first contribution to the Allied cause raised nine regiments of engineers. Though almost 60 years old, Parsons volunteered and assumed command of the Eleventh Engineers Regiment of the First Army." As a founder of the Military Reserve, he had marched the previous summer at the head of the engineers' contingent in the Great Preparedness Parade.

Klapp joined too, and like Parsons was commissioned a major; but instead of being given a command, he was sent to Fort Belvoir near Washington where with other middle-aged executives "he was harangued by a young second lieutenant from West Point to suck up that tremendous gut." Douglas "wanted desperately to go," but the Army was unwilling to give him a rank adequate in pay to support his wife and children. However, they offered to make him a colonel as engineer of maintenance of the Panama Canal. He modestly refused the rank "because it was not a military duty," but accepted the job.

While the Eleventh Engineers trained at Fort Totten, New York, Parsons was dispatched to Europe as chairman of a three-man commission to assess the dimensions of the logistical task. He found it awesome—no less than the care and provisioning of an entire new army, whose engineering and auxiliary troops alone comprised 100,000 men:

Not only were the men to be sent across the ocean, but also all their supplies of every nature, arms, ammunition, clothing, food. To permit the landing of the men and their supplies, there must be berthing places for the ships. But these berthing places did not exist and the material for them, the piles and timbers, were probably still standing in American forests. When the piles and stringers for the wharves had been felled and sawed to size, had been sent across the ocean and erected into wharves, there were no camps for the soldiers to move into or storage buildings to house the perishable supplies. After landing the men and supplies, they could not be moved from base ports until locomotives and cars should be sent from America, because France no longer possessed enough rolling stock to meet its own needs. Nor could France equip the trains thus provided with trained crews, as there was no surplus of man power. Locomotive- and train-men, like the locomotives and the cars, must come from overseas, and, finally, the very rails must be manufactured and sent abroad to permit the moving of the trains from the seaboard to the front.

It was an appalling picture, but . . . if the commission erred in judgment, it was in underestimating . . . the requirements.

Nevertheless, Parsons was exhilarated by "the great part there was for the engineers to take," and as he hurried about London, "saluted at every step with free access to the War Office," the importance of his own position seemed almost to go to his head: "The French gave us a royal reception," he wrote to his daughter on June 2, 1917. "I have been received by the Minister of War and other officers, and everything is at my disposal. But the culmination was yesterday, when I think I had the greatest day of my life. General Pétain, who is now commander in chief of all the forces in France . . . invited me to his headquarters at Compiègne for luncheon. . . . It was an extraordinary scene and very hard for me to realize that I was sitting alongside men who control the great combined army of England and France."

On June 14 Parsons' regiment sailed on the *Carpathia* for England. He linked up with them at Plymouth Harbor, and from there they traveled by rail to Folkestone, crossed the Channel to Boulogne, and on August 18 arrived at the front on the Somme near Peronne — "the very first American forces in France."

Under Parsons, the regiment, known as the "Fighting Engineers," became a legend. "From the Atlantic to the Vosges, from the Mediterranean to Flanders, they built roads, railroads, bridges, docks, and warehouses, and dug and held trenches." In the final days of the war, when the battle formed "one continuous struggle all the way from the Vosges Mountains to the North Sea," their efforts (according to General George R. Spaulding) contributed decisively to the Allied success.

In all the Eleventh took part in five major engagements, for which they received five silver bars on the staff of the regimental colors. Parsons himself was awarded the Distinguished Service Medal, the Victory Medal and five clasps, the Distinguished Service Order of Britain, the Office of the Legion of Honor of France, and the Order of the Crown of Belgium.

But his paths of glory were darkened by many a grave. Though in numerous letters home he zestfully chronicled his own advancement through the ranks (from major to lieutenant colonel to full colonel to his brief attachment to General Pershing's personal staff), he also sought to convey something of the "culminating horror" of the war.

"The singular thing about a modern battlefield," he wrote ruefully, "is that no life is visible." At Peronne (June 3, 1918), once a city of 10,000, he found "not a single house intact," and where there had once been beautiful villages and farms, the ground was so torn up by shellfire "that in places huge craters were actually touching and for miles the surface of the ground was gone." A few weeks later he arrived at Verdun. "You have been to Pompeii, have you not?" he wrote his daughter. "If so, you have seen a dead ancient city. Verdun is a dead modern city, with heaps of dry bones, German and French, all jumbled in the debris."

He thought his little granddaughter had it right when in a misspelling in a letter she spoke of the "hole world."

The war ended at the eleventh hour of the eleventh day of the eleventh month in 1918. In Paris, he said, "the people were very restrained. They have suffered too much, have lost too much to be gay."

§

The rarest book Parsons owned (in his rare and wonderful library) was *Ancient Military Art and Science,* by Roberto Valturio, published in 1483. In all probability, Parsons knew as much about the history of his profession as any man of his time; and therefore both as a profound scholar and a living witness he was uniquely equipped to assess its metamorphosis under the impact of the war.

"There was a time," he reflected in 1919, "when engineers were exclusively military men, when the great pieces of construction other than in architecture . . . were fortifications; while the only intricate types of machines . . . were engines of war." The Industrial Revolution, however, chiefly through the application of steam as a source of power, had thrust civil engineering to the fore. Canals, harbors, lighthouses, roads, and other works, "dissociated from any military connection, began to assume a more ambitious character so as to approach in magnitude the works of national defense." As a result, military engineering adopted as its special province the unassailable fortification, immobile gun emplacements, and impregnable walls.

But at the front in 1917, Parsons saw, with all the men that fell, that old way fall. It "lay buried under the ruins of Namur and Maubeuge, where massiveness, immobility, and heavy castings were smashed, cracked, and crumbled by a few shots from the German 420 mm. guns." Then, Phoenix-like and with a vengeance, military engineering arose from its own ashes. Almost overnight it became "advanced civil engineering," for in addition to bastions, counterscarps, and tenailles, and the rudiments of roads, bridges, and surveying, it embraced electricity, bacteriology, chemistry, metallurgy, geology, and physics. Indeed, there was now "almost nothing in the whole range of applied science that the military engineer should not know something of, and in much of it he must be expert."

§

After a year of assessing the war damage in Belgium, Parsons came home in the summer of 1919 and was promoted to brigadier general in the Reserves. He was thus a general twice over, for he was already a general in the National Guard. Nevertheless, as he had told Alfred Noble sincerely in his letter of 1905, he preferred to serve in the ranks. He wrote to a friend: "Colonel is like 'father' in a family. It is the highest rank you can have and still have contact with the men."

While Douglas had been in Panama, and Klapp* and Parsons at the front, Brinckerhoff had managed the remnant of the firm. "Woefully undermanned and virtually alone" except for Thoresen, he tried to keep in touch with former clients and scare up some that were new. A few of the contracts he secured, in keeping with his own orientation, were in rapid transit (studies for Philadelphia, Chicago, and Cleveland); but most were for the war effort and probably reflected Thoresen's training as a marine engineer: two graving docks for the

---

*Klapp had eventually joined Parsons in France.

Navy at Portsmouth, drydocks at Norfolk and Newport News, ferry slips and terminals at Portsmouth and Norfolk, and five ferro-silicon plants in New York, Maine, and Quebec.

The drydocks at Newport News were a great success, and in 1919 King Albert and Queen Elizabeth of Belgium christened them officially on a state visit. But one of the ferry slips at Norfolk went awry. Though the coordinates were carefully established by an employee of the Corps of Engineers, the surveyor had made an error, and the slip was too narrow for a ferry to get in. When Douglas returned from Panama and reviewed all the work at hand, he hired seven tugboats with steel hawsers and pulled the jaws of the slip apart.

The firm, in fact, had just been hanging on. In 1918, there were 16 names on a payroll that averaged about $4500 a month.

The 16 sweated for their keep in more ways than one. Because coal was in short supply, the government instituted "coal-less Mondays" when all public buildings, schools, and many businesses were closed, and trains operated at a reduced speed or deleted scheduled runs. Office buildings that remained open had to turn their heat down and shut their elevators off. Unfortunately for the staff of Barclay Parsons & Klapp, the firm at 60 Wall Street was located on the 25th floor.

Each Monday morning during the severe winter of 1917-18, the climbers gathered in the lobby weighed down with layers of heavy clothing and carrying their lunches in boxes, pails, and bags. After a long and exhausting single-file march to the top, work began dimly under ten-watt energy-saving light bulbs and a cringing dread of having to make an errand run.

This regimen continued during the long, sweltering summer until the fall when the Armistice was signed.

Meanwhile, the military utility of the Cape Cod Canal was borne out. In May 1918 German U-boats were sighted off Martha's Vineyard, and after one of them surfaced and fired on the tugboat the *Perth Amboy* and sank four of her barges, the government seized the canal as a war measure so "traffic could proceed without molestation by the enemy." At the close of the war, however,

the government coolly attempted to turn back the canal without any accounting for tolls or expenses incurred in its operation. The owners refused to receive it, and for two years it was run by an association of employees under the supervision of the firm. The government and the Canal Company finally decided the government should take over the canal at a price to be fixed by the chief of the Corps of Engineers as arbitrator. Congress refused to appropriate for this purpose more than a sum of two million dollars less than the judgment of the arbitrator, and the Canal Company was finally forced to accept.

This was bad for the stockholders but good for the canal, which for lack of income had been poorly maintained. Accordingly, the Corps took it over on March 31, 1928, and began a series of major renovations and repairs. The deteriorating banks were reinforced; it was deepened to 30 feet, widened to 500, and significantly lengthened by straightening the western approach channel 4.5 miles up the middle of Buzzards Bay. The Corps also ran it toll-free.

Thereafter it attracted a heavy traffic and thus became—and has remained—the widest artificial waterway in the world.

CHAPTER 8

The men reassembled; Parsons' professional family grew. Brinckerhoff and Douglas were admitted to partnership, their names added to the title of the firm, and in November 1919 they moved to new quarters at 84 Pine Street which Klapp decorated with antiques he scavenged from abandoned old Bowery brownstones.

In 1920 Parsons resumed his prewar focus with his most ambitious hydroelectric project yet—the Sherman Island Powerhouse and Dam. He put Douglas in charge, enlisted his brother Harry as a consultant, and hired Alden Foster, a specialist in hydraulics, and Sanford Apt, an expert in the hydroelectric field.

Actually, the power plant and dam were not supposed to be feasible, which for Parsons, at least, undoubtedly gave the project an extra appeal.

Located 8.5 miles from Glens Falls just above Sherman Island on the Hudson, the dam had to be thrown across a narrow gorge with relatively steep slopes on either side. The gorge was made, for the most part, of fine yellow pervious sand and compacted boulders of Adirondack gneiss. But the gneiss was all on one side. The challenge, therefore, was to build a substantial multiple-arch, reinforced-concrete dam and powerhouse in such a way that it would settle equally in unequal ground. Douglas thought he could do it; others thought not; and when he presented his plan at a meeting of the American Society of Civil Engineers, one distinguished colleague stood up and shouted: "I'll stand on the banks of the Hudson and watch your plant float by!"

The solution, of course, was to equalize the ground, which Douglas did by inserting a slab-like concrete cut-off wall in the sand on one side, and on the other, sand fill between the bottom of the dam and the gneiss. In other words, he made the depth of sand to the rock and the wall the same.

Completed in 1923 for the International Paper Company, the plant and dam still stand.

Meanwhile, in related activity, the firm had become involved in a comprehensive Water Power Development Investigation sponsored by the utility companies of New York State. Although a government project, the investigation was managed out of Parsons' office by John P. Hogan, a member of the Board of Water Supply. Hogan's personal and professional ties to the firm were strong. He had been Klapp's assistant on the IRT, and afterwards on the Canarsie Extension of the Brooklyn Heights Railway Company, and had served with Parsons and Klapp in World War I. In his promotion from captain to lieutenant colonel, he was almost as heavily decorated as Parsons himself, with the Distinguished Service Medal, the Office of the Legion of Honor, the Order of the Purple Heart, and the Conspicuous Service Cross. As an expert in water power and supply, he had worked from 1906 to 1913 on the Catskill Aqueduct, the Schoharie and Ashokan Dams, and on New York City's deep underground tunnels and surface reservoirs.

The investigation Hogan directed was far-reaching, and concentrated on the development of the Adirondack Plateau. There, where rivers flow away in all directions, he thought a number of large lakes might be created or enlarged and, once interconnected, their water diverted at will to any of a number of plants during periods of greatest demand. Moreover, this peak power could be combined with the continuous power of the St. Lawrence to meet the varying daily needs of the electric market.

Hogan's report was convincing. In the summer of 1921 the Frontier Corporation (an alliance of General Electric, the Aluminum Company of America, and Dupont) asked Parsons to negotiate a cooperative arrangement for this development with Canada and other interested parties, and in the following spring enabling legislation was signed by Governor Miller of New York.

Miller, however, was up for reelection against Al Smith, and Smith promised, "on the grounds that the power be conserved for the people," that if elected he would have the Miller bills repealed. He was elected, and, said Hogan, "the development of the St. Lawrence River was postponed for 30 years."

Hogan was bitter about this to the end of his days. He claimed that the Adirondacks were inadequately developed by private interests, while "the people did nothing."* As for the St. Lawrence itself, "when

---

*All quotations from Hogan are from his unpublished *Memoir.* See bibliography.

we consider," he wrote in the mid-1950s, "that it is costing four times as much as it would have cost at that time to construct the development, and that the people have lost over 100 million kilowatt hours of energy which would have cost about one-third as much as it will now, one fails to see where they have benefited. The great industries . . . have gone elsewhere, as well as the numerous other manufacturing interests that cheap power would have attracted to the state."

In any case, at the conclusion of the investigation in 1923, Hogan had joined the firm.

Hogan, said a colleague, was "a man to be feared." Small, spry, alert but irascible, with "a grand sense of humor" and "a lovely aroma of fine old bourbon about him," in the morning he might appear the merry Irishman, but after a couple of bourbons at lunch he could be rough. "I went down to his office the first day I was there," remembers Helen Yasso, his feisty confidential secretary for many years,

> and he was rustling papers in the basket and hanging out the window and other things, and he turned around and glared at me and said, "You mean to say you're sitting there and you didn't get one word of what I said?" And I said, "Well, look, Sir, if you will turn around and face me so I can hear you, stop rustling papers, then I'll get every word." "Well, you had better," he said, "because if you don't you're fired." And from then on we were pals. He wouldn't bully me and I respected him. He turned out to be a wonderful man to work for.

But she had to keep him in line. He was a bundle of nerves, with a number of odd and disconcerting habits including one, while dictating, of zipping his fly up and down. "Colonel," she'd snap, bringing him sharply to, "adjust!"

In 1926 Hogan was made a partner, and between 1923 and 1928 was responsible for the design of a peak power plant for the New York Power and Light Company, a Hudson Falls plant for the Union Bag & Paper Company, and the expansion of the Spier Falls Power Plant that Parsons and his brother had designed in 1900. He also acted as consultant to the Cohoes Power & Light Corporation and to the State of New Hampshire on the redesign of its dams.

But though the hydroelectric work came flowing in, the overall postwar economic climate was cautious, and for consulting engineers at least, new terms were established for the way certain business was done. Financial backers of projects took the unprecedented step of ensuring a return on their investment by demanding that engineers guarantee construction costs, particularly in new classes of work. The only way an engineer could do this, of course, was by constructing the project himself or (barring that) by keeping the contracting end of it under his strict control. To meet this new demand, and to ensure the firm's continued growth, Douglas improvised the Parklap Construction Companies (Parklap = Parsons + Klapp) as independent entities but operating in effect as the firm's construction arm. The device assumed two forms: the Parklap Construction Corporation, to undertake heavy construction; and Parklap, Inc., for the construction of buildings.

"The market for civil engineering is ever different," observed Douglas. "A firm like ours must change in order to survive." It was the Parklap device, as well as Douglas' remarkable financial acumen, that led Hogan to describe him as "the driving force of the partnership and the second founder of the firm."

Indeed, the Parklap companies were fabulously successful, and between 1920 and 1939 completed more than $50 million worth of work, including railroad and harbor terminals in Toronto, Montreal, Buffalo, Albany, and Detroit; railroad car manufacturing plants in Ontario and oil storage tanks in Quebec; numerous houses in Florida, as well as a recreation pier and the Don Ce Sar Hotel in St. Petersburg, a yacht basin for the Vinoy Park Hotel, the Tampa Union Terminal, the St. Petersburg News Building, bulkheads and wharves at Belleair, St. Petersburg, and Fort Pierce, and the first large-scale plant for processing orange juice.

Though Klapp ran the main office of Parklap, Inc., which he had opened in Florida in 1924, during one interlude he went to Madrid to build a towering skyscraper for IT&T. For some obscure reason (the city is not in a seismic zone) he was told to make it earthquake-proof, and though never tested by a quake, it survived something comparable when in a merciless bombardment during the Spanish Civil War it was hit over 20 times.

When he returned to Florida, Klapp set up residence for himself in a little Cuban-style house in St. Petersburg, into which he incorporated antique tiles he had collected while in Spain. Much to his chagrin, however, the plaster between the tiles lacked "that certain patina which beautifully sets them off"—because, of course, that usually comes with age. One day a hurricane swept through the Florida Keys and blew in the rain. When the house dried out, behold the patina was there, and the Architectural Society gave his house its award of the year.

Meanwhile Douglas enjoyed an interlude of his own to work on what he regarded as the crown of his career. Between 1925 and 1932 with the architectural firm of McKim, Mead & White and the Corps of Engineers, he collaborated on the design and construction of the Arlington Memorial Bridge in Washington, D.C. First suggested by Andrew Jackson as a symbolic link between North and South, the bridge crosses the Potomac in a direct line from the Lincoln Memorial to Arlington Cemetery, and therefore occupies what is probably the most prominent bridge site in the United States. Stretching some 2100 feet between pylons, with eight reinforced concrete arches and a center double-leaf bascule span, its 216-foot steel draw span is one of the longest, heaviest, yet swiftest in the world. The bridge's finely wrought yet majestic character, variously decorated with sculpted eagles and bison, and dressed in pink granite ashlar from North Carolina, befits its status as a national monument.

No expense was to be spared, though Douglas was determined to economize. He insisted on the pink granite over the more expensive New Hampshire gray, and on machine-tooling the facing which the architects had wanted done by hand. "He got them into his office with numerous samples of both, and they were unable to tell the difference at a distance of twenty feet." Though perhaps not the best aesthetic test to apply, with $2 million at stake his demonstration was hard to refute.

In 1927, with Parklap in the wings, Douglas negotiated a contract for a work of only slightly less renown—a vehicular tunnel under the Detroit River between Detroit, Michigan, and Windsor, Ontario, Canada. A number of bankers were involved in the financing, and partly to adjudicate the disposition of the bills, Thoresen immediately went to Windsor to figure out, with the help of a theodolite, where the exact dividing line between the United States and Canada lay.

The construction was divided into three sections: the open approaches, which were trenches cut in the dry; tunneling by shield from the approaches to the river; and the tunnel under the river, which was built by sunken tube.

The shield was a cast-iron circular machine with a cutting edge thrust forward by 30 hydraulic jacks. Before each thrust, workers sliced away at the dirt from within with power-operated knives; after each thrust, a hydraulically operated erector arm picked up segments of the welded steel plate lining and swung them into place behind the shield to form a ring. The result was a continuous enveloping steel cylinder with a diameter of 32 feet.

The central river section was made up of nine steel tubes, each weighing about 800,000 pounds. Built on dry land, they were launched like ships, floated into position, and sunk in a trench dredged in the bottom of the river. To manage their exact alignment, masts were attached to each and matched in a line from shore to shore, then the tubes were welded together by divers and attached at either end to the sections built by shield.

The overall error in alignment was less than one inch.

In the course of preparing the trench, three corpses were dredged up, along with some firearms, an old wooden waterpipe, and an Indian sling with a cord of bark fiber still attached. More notably, the mile-and-a-half long tunnel, completed in 1930, was the first subaqueous tunnel to connect two countries, and the first long vehicular tunnel constructed by the sunken tube method anywhere in the world. It also marked the greatest use of structural welding in any construction project up to the Second World War.

While the Detroit-Windsor Tunnel had been pressing forward, the American economy had been tottering towards the crash. Despite ominous signs, opportunity appeared rife, and many lingered in suspiciously remunerative investments for one last profit too long. Douglas, however, appears to have had his nose to the wind. In 1927, he "smelled a rat" on a transit job with which Brinckerhoff had been busy in Cleveland, where he had been hired by two local bankers, the Van Sweringen brothers, to develop a portion of their Nickel Plate Railroad as a rapid transit line. Brinckerhoff had assembled a large work force, and the expenses were running about $20,000 a month. But then a bill was inexplicably returned with various items struck out. Douglas settled the account at once and withdrew. A few months later the Van Sweringen holdings collapsed.

In Florida, where the housing boom had given Parklap, Inc. just about as much work as it could handle, the risks were more seductive. Then a bank closed here, a company failed there, and Douglas advised Klapp to wrap up all remaining contracts, which in a most timely fashion he did. Before the end of the year, millions of dollars of stock had become worthless.

By and large, the Crash of 1929 put an end to heavy construction opportunities, at least in the United States, and after completion of the Detroit-Windsor Tunnel the Parklap Construction Corporation was dissolved in all but name. Parklap, Inc. might have gone the same way had it not been for the resourcefulness of Gene W. Hall, who as Klapp's chief assistant in Florida had handled the work from the contracting end.

Hall had come into the firm in 1915 as an office boy. Physically large with a big neck and piercing eyes, he looked like a successful contractor, with his ruddy face and powerful jaw. Though he had no formal education to speak of, he had "a mind like a bear trap—once something got into his head it was *there.*" In 1929, what was there was a knowledge of construction, and as he had made no other significant contribution to the firm thus far, he was loathe to let it go. Hall took hold of Parklap, Inc. despite Douglas' initial misgivings, and developed a new and profitable business advising insurance companies on their contracts and completing contracts in default. After a law to stimulate housing was passed in 1933, he organized Parklap National Builders, Inc. to finance and construct middle-income housing, some of it co-sponsored by the government and private industry: near Linden, New Jersey, for General Motors; near Dundalk, Maryland, for Bethlehem Steel; and near Kingsport, Tennessee, for Eastman Kodak. Under Thoresen's favoring eye, he also helped build two Christian Science Churches in New Jersey, in Linden and Haddon Heights.

Meanwhile, the success of the Detroit-Windsor Tunnel had led in 1930 directly to the design and construction of a comparable tunnel under the River Scheldt in Antwerp, Belgium. It may be said that together these two substantial projects sustained the firm during the worst of the Depression years.

The Scheldt Tunnel contract had been won against stiff competition. French, British, Dutch, Belgian, and German concerns had also submitted bids, but by comparison the firm's own estimates of time and cost were so low that the Belgian government demanded guarantees. Douglas, by now quite intimate with this requirement and a past master at meeting its terms, agreed; but he insisted in return that American methods (on which the estimates were based) be written into the deal. Accordingly, the Parklap Construction Corporation (in its last revival) was brought in to form an ad hoc partnership with Pieux Franki, the Belgian contractor, with Parklap furnishing the controlling personnel: superintendents, shift bosses, master mechanics, and others.

To celebrate this arrangement, a party was held in the executive offices of Pieux Franki. Thoresen, as the tunnel's designer, was naturally invited, and he brought along his eighteen-year-old-son. The president of the company buzzed his secretary, who opened a safe and produced a bottle of vintage Napoleon brandy about a hundred years old. He set glasses before each of his guests, and ebulliently pulled the cork. But before he could pour out a drop, Thoresen—never so dry—inverted his glass, and quickly reached over and inverted the glass of his son.

The purpose of the tunnel (about the length of the Detroit-Windsor), was to open up the west bank of the river to development. A bridge had been ruled out as an interference to shipping, but so had a tunnel by sunken tube, as the quicksand bed of the river could not retain a trench. On the other hand, there was a permanent clay stratum 86 feet down which was hard and impervious and, using the shield-and-compressed-air method, ideal for tunneling through.

As with the Detroit-Windsor Tunnel, the riverside approaches were conventional open-cut, dug rapidly with a drag line and clamshell bucket; but it took four months to get the shield started because the ventilation shafts (through which it would have to pass) could not easily be sunk in the runny, waterlogged ground. That ground had first to be frozen, which was done with injections of brine. The brine passed through a freezing circuit of pipes, and eventually turned the surrounding ground into a solid wall.

In late October 1931, the shield began driving toward the river from the west.

"American methods," however, did not at first sit well with the Belgians, and the early labor turnover was 400 percent. "After one ring of two and a half feet had been installed," recalled Hogan, "the workmen called it a day." But when wages were raised and a bonus system instituted, production more than quadrupled, from 3 to 13 rings per 24-hour day.

To protect the workers in the compressed-air caissons from "diver's palsy" or the "bends," strict discipline was maintained on entering and leaving the locks. Close to the shore, the air pressure requirements were reduced by sinking a series of well-points which reduced the hydrostatic pressure above; under the river, the pressure was adjusted to synchronize with the tides. When it exceeded 30 pounds per square inch, the shifts were split so that no man remained below for more than four hours. As a result, the accident rate was virtually nil.

Such methods paid off. Over the course of 400 days, an average of four rings or ten feet were completed per shift, and on February 28, 1932, the shield holed through to the opposite bank.

The tunnel was an impressive achievement. Though a mile and a half long, it had taken just eighteen months to build, as against the best European estimate of four years; moreover, it had cost $2.5 million below the next lowest bid. Even so, the margin of profit for the firm was handsome.

(And the tunnel was sound. During the Second World War, it was blown up twice—first by the Belgians retreating south, later by the Germans retreating north. On both occasions, a truck loaded with explosives was driven into the middle of it and detonated; and on both occasions, it was effectively repaired.)

For its efforts, the firm earned a $25,000 bonus, and Thoresen and Douglas were knighted by the King of Belgium.

There was also a smaller tunnel for pedestrians and bicyclists, which Douglas had tried to get but which the Belgians, "when they saw how easy it was," took on themselves. However, hardly had they burrowed more than 600 feet (still under land) when the tunnel collapsed, and the firm was enlisted to recover it for them at a cost of $400,000.

Nevertheless, the Depression hit the firm hard, and in January 1933, Douglas concluded that as soon as Roosevelt was sworn in on March 4th, he would take some sort of drastic action which would adversely affect engineering and construction for some time. On February 22, when the office was empty, he called Klapp, Brinckerhoff, and Hogan into conference and said he believed a bank moratorium would be declared. Accordingly, he proposed that the four partners reduce their drawing account from \$12,000 to \$4,200 a year; that \$40,000 plus three months' estimated payroll be withdrawn from the bank and placed in a safe deposit box; and that, while he couldn't take a holiday himself, the others should take six-week vacations and withdraw from their personal accounts whatever cash they could spare.

CHAPTER 9

The moratorium, of course, was declared as predicted; and in the ensuing months one of the things Douglas did to try to raise money was to advise banks on construction loans. In this he may have taken his cue from Hall, whose advice to insurance companies had brought in a tidy sum. However, the banks usually rebuffed him with: "We don't need technical assistance in making loans," to which Douglas usually rejoined, "We never had to close our doors."

Strictly speaking, that was true; though as an economy measure in September 1929 they had closed their doors at 60 Wall Street and moved to a little waterfront brownstone Parsons owned on Maiden Lane. Before its conversion to an office building, 142 Maiden Lane had served as a carriage stable.

It was a quaint place. A library was established on the first floor "with subdued lighting and heavy paneling," and a dumbwaiter ran behind the stairwell with openings on each floor. As there was no elevator, the dumbwaiter saved a lot of running up and down, though on occasion errant mail was rediscovered at the bottom of the shaft. Other space on the first floor was rented out to Waddell & Hardesty, bridge engineers, and one room to a two-man sanitary engineering team run by a Mr. Ketchum, "who was known for his ability to make a long-distance call without use of a telephone." Waddell, when he had nothing else to do, liked to give speeches on fishing.

The work force of Parsons Brinckerhoff—about 25 engineers and their assistants—occupied the third and fourth floors with the partners on the second. The basement, which was generally full of water, held the files. "Every time the tide rose in the East River," remembers one veteran, "the basement would flood. We had little duckboards down there to walk on, and the cabinets sat on duckboards, too. The files were always damp, and some of the old cloth tracings which were made of starch on pure linen exuded an odor I will not forget. Sometimes we had to dredge them up. Fortunately there were no animals."

Nearby was a coffee mill, which discharged the hulls from roasted beans from stacks on the roof. In the summer, downdrafts brought them in through the open windows and scattered them across the drawing boards and into everyone's hair. From another side, the olfactory delights of the Fulton Fish Market came floating in.

Parsons had originally bought the building as a real estate investment because he thought the proposed Second Avenue subway would one day build a station nearby. And it was there that he spent his last working days.

Though he had officially retired in 1924 at the age of 65, on most afternoons he could still be found in his office where, as the *pater emeritus* of his family of engineers, he kept himself abreast of their work. Douglas, who had assumed the day-to-day direction of the firm, always deferred to him as a matter of course.

Parsons had other interests too, as he had always had, to which he now devoted more time. One was the archaeology of Central America, which since 1900 he had pursued as a trustee of the Carnegie Institution of Washington. Over the years he had encouraged, guided, and occasionally (as warranted) criticized the Institution's work. "Although the Warrior volumes are magnificent," he wrote its president in June 1931, "they lack proper recognition of the careful way in which the excavation was carried out; nor is there a picture of the Temple of the Warriors as it stood before any work was begun. This I regret because the transformation from a rough tree-covered hill to a stately building is startling." Sometimes a suggestion of his own led to a discovery in the field, as when a secret cache of artifacts was found beneath a Mayan altar. He was, indeed, "in constant touch with every aspect of the archaeological program," and "the magnificent Mayan ruins preserved in Guatemala and the Yucatan, along with the knowledge recorded in the classic publications of the Institution, owe something to his early initiative and care."

There was also his literary and historical writing, which had accumulated over the years: numerous papers and speeches, and (besides his two early technical tracts) four books: *Rapid Transit in Foreign Cities* (1895), *An American Engineer in China* (1900), *The American Engineers in France* (1920), and *Robert Fulton and the Submarine* (1923). In addition, he had been laboring away at what would prove to be his posthumous historical masterpiece, *Engineers and Engineer-*

*ing in the Renaissance,** which in its field remains unsurpassed. Over the course of 20 years, as time and opportunity allowed, he had gathered material for this volume at the British Museum, the Bibliothèque Nationale in Paris, and the Vatican Library. The result was a 661-page compendium of notable achievements in all branches of engineering from 1450 to 1600.

Its "hero," not surprisingly, was Leonardo da Vinci, whose birthday Parsons coincidentally shared—though Parsons could not have known this, since the date was first determined by a scholar in 1939.

In the course of his research, Parsons had found the Vatican Library almost impossible to use. So he suggested to the Carnegie Endowment for International Peace that, "as a real contribution to international peace and understanding," it help subsidize its modernization. The Endowment agreed, as did Pope Pius XI, and in the fall of 1927 the invaluable work was begun.

With all his interests and activities in retirement, Parsons gave aid to the firm as he could. As late as August 1931, almost a half-century after he had first hung out his shingle as consulting engineer, he was still actively soliciting business, as reflected in a letter to William T. Cosgrove, President of the Irish Free State.

§

Parsons died at the age of 73 on May 9, 1932 from a pulmonary embolism that developed after surgery on an infected arm. Funeral services were held at Trinity Church, where he had been a vestryman and warden, and his coffin was draped with the regimental flag of the Eleventh Engineers. A motorcycle escort led the funeral cortege through the city, and with many dignitaries attending he was buried at All Saints Cemetery in Navesink, New Jersey.

Preferring ultimately to serve in the ranks, he is identified as a colonel on the headstone of his grave.

Nicholas Murray Butler, Parsons' undergraduate companion, president of Columbia University, and possibly his closest friend, said of him: "He was a true representative of the culture and refinement of

---

*See the original edition (William & Wilkins Co., 1939) or its 1967 reprint. The M.I.T. edition (1968; paperback, 1976) injudiciously omits the first two chapters, and substitutes an unfocused preface for Nicholas Murray Butler's sensitive introductory remarks.

old New York, and his interest in education, in religion, philanthropy, and in public service all came as naturally to him as did the ordinary incidents of life." Parsons' ties to Columbia were particularly strong. It was his alma mater; he had served on its Board of Trustees since 1901 (and as chairman since 1917), and as chairman of the Joint Administrative Board of the University's Medical Center and hospital for which he had driven the final rivet on May 24, 1926.

On October 1, 1934, a tablet in his memory was unveiled at the university's St. Paul's Chapel. Butler said on that occasion: "General Parsons had certain qualities that were all his own, and which gave him his just and commanding influence in the life of Columbia. He never lost his poise; no problem was too sudden or too difficult to find him in so emotional a state that he could not approach it justly and wisely and with clearly opened eyes. He had unbreakable courage. He was not afraid of a difficulty; he could look beyond to its solution. He had that fine feeling and kindly courtesy of the cultivated American gentleman which meant so much in the life of all those with whom he was brought in contact. His name will be remembered so long as this University may last."

UPON A SURE FOUNDATION:
1885–1932

*William Barclay Parsons as a young man, soon after graduating from the School of
Mines, now Columbia University's School of Engineering, in 1882.*

The Engineering Department in the School of Mines.

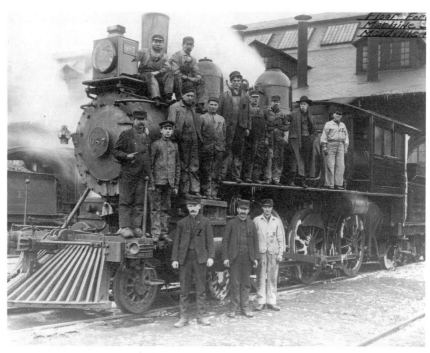

Erie Railroad Repair Shop, ca. 1885. Parsons' first job was with the Erie, where he rose quickly in the ranks and in 1884 was put in charge of repairs on the Greenwood Lake Railway.

Parsons in China, 1898. He surveyed a railroad route between Hankow and Canton, a stretch of a thousand miles including the unexplored province of Hunan.

Ground breaking for New York City's first subway, the IRT, March 24, 1900. As Chief Engineer for the project, William Barclay Parsons ceremoniously wielded the pick.

Cut-and-cover construction at New York City's Union Square.

Construction on the Steinway Tunnel, a subaqueous tunnel between Manhattan and Queens that extended the subway to Long Island City. To speed the work, Parsons created a large working platform in mid-stream on Man-o-War Rock, visible in the background.

At work on the Steinway Tunnel, ca. 1906. The tunnel, now known as the Queensboro, was completed in 1907.

Dignitaries inspect the IRT on opening day—October 27, 1904. The City Hall station, with its skylights, vaulted ceilings, and intricate tiling, was the subway's most ornate.

Eugene A. Klapp joined Parsons' fledgling firm in 1905. By 1909 the firm was known as Barclay Parsons and Klapp. Klapp had been chief engineer for the northernmost division of the New York City subway and was Parsons' deputy chief engineer for the Steinway Tunnel. Klapp developed work in Florida and Cuba for the firm.

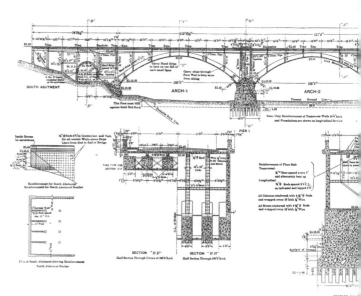

The Machina and Passenger Landing.

Site of Dock Nº 3 in front of Machina.

Projects included a study of the port of Havana, Cuba (above) and design of the Almendares Bridge (below) also in Havana, the first concrete arch bridge in Latin America.

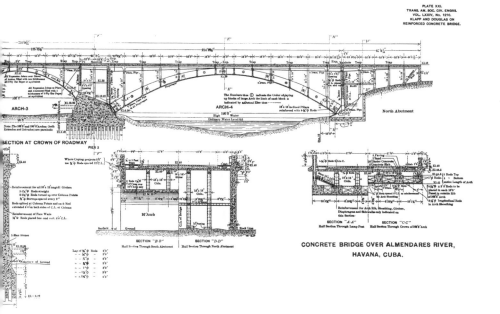

PLATE XXI.
TRANS. AM. SOC. CIV. ENGRS.
VOL. LXXIV, No. 1210.
KLAPP AND DOUGLAS ON
REINFORCED CONCRETE BRIDGE.

CONCRETE BRIDGE OVER ALMENDARES RIVER, HAVANA, CUBA.

Parsons hired Henry M. Brinckerhoff, a leading traction power engineer from Chicago, in 1906. Brinckerhoff shared the patent for the third rail.

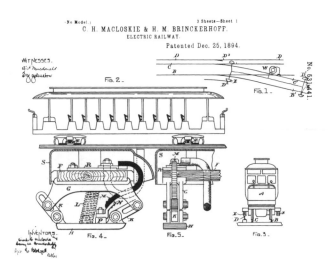

Drawing from patent for the Electric Railway, Dec. 25, 1894

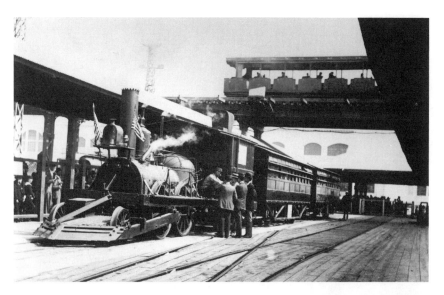

The 3.5 mile Intramural Railway (elevated line above locomotive) demonstrated Brinckerhoff's third rail and was among the technological marvels at the Columbian Exposition of 1893 in Chicago.

A dredge at work on the thirteen-mile-long Cape Cod Canal. The canal was completed in July 1914.

Cape Cod Canal Company President August Belmont and Chief Engineer William Barclay Parsons shake hands across the stream which made Cape Cod an island.

Parsons served as a colonel in World War I. His "fighting engineers," the Eleventh, became a legend and Parsons was made a Brigadier General when he entered the Reserves at the end of the War.

Among the firm's early projects was a transportation study for the United Railways of Yucatan which ran to the port of Progreso. In his later years Parsons encouraged the Carnegie Institution of Washington to excavate the Mayan ruins in Yucatan.

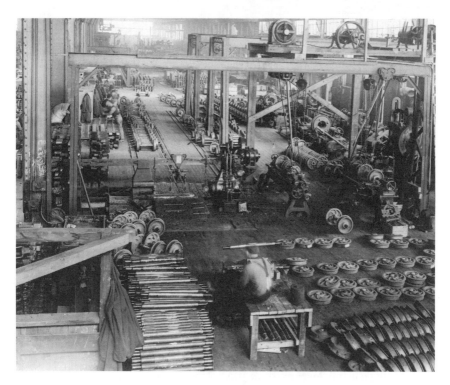

A railcar construction shop, the National Steel Car Company, Hamilton, Ontario. In 1919 the firm provided designs for expansion of the shop.

The Spier Falls Power Plant in New York State, designed by Parsons and his brother in 1900 and expanded by the firm in 1926.

*Walter J. Douglas joined the firm in 1908 and was effectively in charge at Parsons'
death in 1932.*

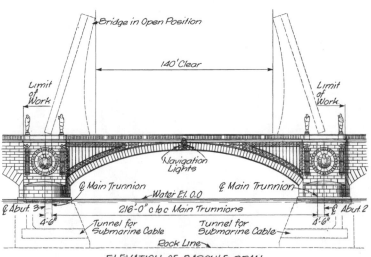

*Center span of the Arlington Memorial Bridge in Washington, D.C., a concrete arch
span across the Potomac completed in 1928. Walter J. Douglas was chief consulting
engineer for the bridge.*

Douglas, lower right, with crew at the Sherman Island Hydroelectric Plant, New York, 1922.

*The Detroit Windsor Tunnel, a mile-long subaqueous tunnel under the Detroit River,
connected the United States and Canada. The prefabricated tubes were sunk into a
trench dredged in the river bottom.*

*Completed in 1930, the Detroit Windsor Tunnel project helped the firm weather the
Depression. A partially submerged tube is shown being towed into position.*

Approaches to the tunnel were constructed by cut-and-cover and by the compressed air shield method.

Soren Thoresen, one of Parsons' first two employees, designed the Detroit-Windsor Tunnel and the Scheldt Tunnel in Belgium, and was also partner in charge of many South American projects.

Drilling equipment used for the 1.5 mile-long vehicular Scheldt Tunnel.

River-bank construction on the Scheldt tunnel. The firm also completed a 2,000 foot-long pedestrian tube.

PUBLIC WORK AND THE WAR: 1932–1945

CHAPTER 10

U ntil the Depression most of the firm's clients had been the major private industries. After the Depression the major client was the government, and the projects were public works. Hogan, in both a general and a particular way, did much to bring this about.

About a month after Roosevelt had declared his "bank holiday," Hogan was dispatched to Washington by the American Society of Civil Engineers (ASCE) to lobby for the Public Works Act. He worked closely with Simon Rifkin in drafting the legislation and organized a core of about a hundred engineers to direct its implementation. Subsequently he undertook a similar task for the National Recovery Administration, or N.R.A. With Frank Voorhees, former president of the American Institute of Architects, he called a meeting of several hundred leading contractors and warned that "if they didn't put a horse in the barn they would probably get a mule." After a stormy debate, the contractors went along. In the following year Hogan was made chairman of the Construction League of America.

That summer Hogan was also selected by the Platte Valley Public Power and Irrigation Authority as overall engineer for a $15-million PWA hydroelectric project at North Platte, Nebraska. It involved construction of a diversion dam, canals, power conduits, storage dams, and a powerhouse. To make use of the highest head possible, water was taken as needed from the diversion dam and run 40 miles through canals to the South Platte River, where it was fed through a siphon tunnel and flowed on to a powerhouse.

Hogan managed to transfer about a third of the firm's personnel to North Platte, where they could earn a full salary.

At about the same time, Hogan was also appointed one of four consulting engineers to the Bonneville Dam in Oregon, a $50-million project run by the Corps of Engineers and the first dam across the Columbia River. Again, he drew on the firm for personnel, and at one point the Corps enlisted the great songwriter Woody Guthrie as an "information consultant." Guthrie carried a guitar that had emblazoned on it "This Machine Kills Fascists," and for a fee of ten dollars each wrote about 30 songs. One, a classic of its kind, begins, "Roll on, Columbia, roll on / Your power is turning our darkness to dawn."

Over the course of several years Hogan commuted to both Bonneville and North Platte once a month by air. As he was also vice-president of the ASCE—an itinerant post—he reckoned that in one year he clocked about 70,000 miles by air and 12,000 miles by rail.

In related activities, he sat on Roosevelt's Special Review Board on the Santee-Cooper dam project in South Carolina, was consultant to the Passamoquoddy project in Maine for harnessing the tides, and helped prepare a report for the development of the Delaware and Susquehanna rivers for water power and supply.

Oddly enough, the firm's conspicuous eminence in the hydroelectric field was about to fade. Hogan could not keep up such a pace indefinitely, and increasingly relied on his principal assistant and partner-to-be, Edward Maloney. When Maloney died suddenly, there was no one suitable to take his place.

Certainly not Douglas, whose industry fairly rivaled Hogan's own, with the Sacandaga River Power Plant at Conklingville, New York; the Pennsylvania Avenue or John Philip Sousa Bridge over the Anacostia River in Washington, D.C.; with his responsibilities as consultant to the East River Drive in New York City, and on foundation problems at LaGuardia Airport; and as chairman of Roosevelt's Board of Consultants for an Atlantic-Gulf ship canal across Florida.

On the other hand, the firm's bridge department—essentially Douglas, after Klapp's retirement in 1931—had been revived.

Early in 1932 Eugene Macdonald, a Waddell & Hardesty employee who didn't care to be idle, had given Parsons a call for advice. Parsons said, "Easy to give, but never follow it. Come on over." So Macdonald went down the hall and up the stairs to meet with Parsons and Douglas. Macdonald pointed out to them that though bridge design had ceased to be a gainful occupation, the situation must eventually change, and they'd be wise to have a Bridge Department when it did. They didn't have one now, but he'd be happy to form one, with two of his colleagues, John O. Bickel and Maurice N. Quade. Parsons and Douglas said, "Fine, but you'll have to bring in your own jobs."

In retrospect this brief, cordial exchange determined the growth and direction of the firm for the next 30 years. Macdonald, Bickel, and Quade proved to be a formidable triumvirate. So much talent had not come into the firm at one time since Thoresen, Brinckerhoff, and Klapp.

Macdonald, by seniority at least, was the first among equals. A tall, dour Scot, scarecrow-like and "not too terribly friendly," he answered the phone with "What's the matter?" and brought his lunch to work in a brown bag. Somewhat like Douglas, whose heir apparent (once removed) he soon became, he was an extremely shrewd businessman with a dry, quick wit.

But until he joined the firm at the age of 42 he had been unable to put down roots.

Macdonald had chosen civil engineering "to be out in the open air." However, his first job was in a subway, which was anything but, so he switched to the Hell Gate Bridge—which, he said, "was airy enough, and complied with the idea I had of living dangerously." But the pay he received hardly complied with the risks, which led him into merchandising for Macy's department store. Then in 1917 he went to war under Parsons in the Eleventh Engineers.

For a decade after the war it was all odd jobs: a bridge in Florida, the design of a plant for Henry Ford, and a stint as an efficiency expert for an axle company where he was shown "a certain method of coordinating design with performance." After one tour of the plant he told his boss: "You know that bushing straightener we designed—it doesn't straighten them, it breaks them." The man replied, "Change the title."

This was about 1920. In 1921, a minor depression hit Detroit and Macdonald found himself pounding the pavement as a reporter for the *Detroit News*. It was the Prohibition Era, and most of his tips came from bootleggers who shuttled back and forth across the river at night.

Thereafter, he went to Canada to work on the design of movable bridges, but as he was not a Canadian national his prospects for advancement were slight. Eventually he showed up at Waddell & Hardesty's door.

At Waddell & Hardesty he began to hone his skills, and before the Depression he collaborated with Bickel and Quade on the Goethals Bridge and Outerbridge Crossing in Staten Island, on bridges across the Hudson at Albany and Troy, and on two bridges over the Niagara River.

Quade's early career was less eventful. He had grown up in rural

Illinois, and before college had worked for a small-town bank. But the job was not exactly staid. Every morning he had to meet the milk train coming down from Chicago, take off a large sack, carry it home, and deliver it to the bank when it opened at 8:00 a.m. "Nowadays," he reflected in the mid-1950s, "I wouldn't have taken that job no matter what it paid." The sack contained the entire town payroll.

At the University of Illinois, Quade studied under Hardy Cross, one of the great teachers of his generation, and in 1926 was hired by Macdonald at Waddell & Hardesty.

Quade revealed himself from the start as an exceptional engineer. He was so clever, in fact, that he sometimes outwitted himself. During construction of the Bonneville Dam, the rising water threatened the Bridge of the Gods, a long cantilever-truss span over the Columbia River upstream. Quade saw a way to save money, and suggested raising the bridge instead of demolishing it. The Corps, intrigued, invited him to make a report, which he did, and more: he showed in detail how to accomplish what he had in mind, in a seesaw operation using timber cribbing supports and hydraulic jacks.

A month later he called the district engineer to discuss a contract for final design and plans. "Contract plans?" inquired the engineer, ingenuously. "What do you mean? We gave your report to the contractor and he's hard at work!"

As for Bickel, he eventually proved to be the most versatile of the three. He had graduated in 1921 as a structural engineer from the Swiss Federal Institute of Technology in Zurich, and in 1923 came to Waddell & Hardesty as a designer of movable bridges. He would go on to excell as a mechanical and electrical engineer, and round out his career as one of the best subaqueous tunnel engineers in the world.

§

At Maiden Lane, since times were difficult for all, the two firms pooled their staffs. In fact, they were so intermixed that at one time Macdonald, Bickel, and Quade represented simultaneously the entire staff of Waddell & Hardesty and the bridge department at Parsons Brinckerhoff. In 1938, when Waddell & Hardesty began to revive and moved uptown, they literally had to sort out who belonged to whom.

Anyway, in 1932 Macdonald, Bickel, and Quade set up shop in an unused drafting room, and for the first few months bided their time checking shop drawings at two dollars each. Then Macdonald, who laconically noted that "a bridge engineer is an engineer who has a contract to design a bridge," secured the contract for a new Buzzards Bay Bridge over Parsons' Cape Cod Canal.

The widening of the canal, which was still going on, meant of course that Parsons' old single-leaf bascule span (which no ship, after all, had "carried away") would have to be replaced. The trio seized the day to show what they could do: they built the longest and—of its kind—one of the most beautiful vertical-lift spans in the world. It weighed 4.2 million pounds, was 544 feet long, made of silicon steel, and rode up and down between twin towers carried by cables that passed over roller-bearing counterweight sheaves. The towers (courtesy of McKim, Mead & White) stood 260 feet tall and resembled light-houses, with illuminated beacon-like finial spheres at the top. The outline of the span was itself enhanced by sequestering the operating machinery out of sight in the towers.

Quade designed the span, Bickel the machinery, and Macdonald handled the contract with the client. In subsequent years, Quade's wife would say: "Every time I go to the Cape and pass that bridge, I silently thank it for feeding us for six months. I also check to see if the huge spheres at the top of it—made of a new substance at the time—are still 'shining forever' as the salesman said they would." And so they are.

The bridge, landmark in every sense, created a magnificent gateway to the canal, and on December 27, 1935, the span was lowered and the first train rolled across.

The trio's next major project together was the Jamestown Bridge over Rhode Island's Narragansett Bay. This was a cantilever highway span, and its utility was suggested by the New England hurricane of 1938, which blew out the ferry service. The bridge had to be designed quickly, in compliance with a WPA program that made disaster relief funds available if they could be used promptly. Macdonald and Quade laid out a simple configuration of 20-odd piers, then riffled their files for a plausible overall design. They came up with the Cooper River Bridge, which Waddell & Hardesty had designed in the 1920s for Charleston, South Carolina. They put a label on it that said, "This is the Jamestown Bridge," with a note that advised, "Details will be furnished later." Many of the details were different, of course, but the general configuration was the same.

Other collaborations were the Passaic River Crossing in Newark, New Jersey, and the St. Georges tied-arch bridge over the Chesapeake and Delaware Canal. The latter, the first of its kind in America, reversed the conventional form of the tied-arch (wherein the arch is the heavy member) and made the tie heavy and stiff, with a graceful, slender arch.

On the Passaic River Bridge, they were taught how serious their profession could be. On the opening day, one of their colleagues, a Russian, was discovered sitting in a rowboat underneath the span. "What are you doing there?" they asked. "It is a Russian custom," the man replied sadly, "for an engineer to be under a bridge he designs when the first load goes across."

Quade and the others stood aside.

CHAPTER 11

I n 1937 Hogan received an offer he couldn't refuse. He wanted to refuse it, he said, because "I foresaw nothing but trouble. But for the sake of the firm, I could not."

Reluctantly, therefore, he became chief engineer of the 1939 World's Fair.

The project for its day was enormous: $35 million in construction costs for the standard buildings alone, plus $90 million in specialized construction for the foreign pavilions, booths, and other structures for exhibitors and concessionaires. Moreover, although the Fair was scheduled to open in two years, the site donated by the city (Flushing Meadows) was virtually a wasteland— two-thirds swamp and lagoon, with a cinder dump fifty feet high at its northern end, and not a tree in sight.

To turn this acreage into a fetching fairground able to accommodate millions of people from all over the world, with all the conveniences they would naturally require, was understandably daunting; and so to get it underway with a rush, Hogan drew once again upon the firm for his engineering core. He put Brinckerhoff in charge of all transportation within, and to and from the grounds; Lester Hammond (who had been a vice-president of Parklap, Inc.) supervised engineering in the field; L. B. Roberts, surveys and borings; Sanford Apt, ventilation, air conditioning and heating; John White, specifications; Douglas's son, Walter S., contracts, extra orders and costs; and G. Gale Dixon, water supply. Douglas himself (with the architect R. M. Shreve) prepared the standard building designs, with an accurate budget, in just three weeks.

To stabilize the waterlogged ground, ash from the dump was spread over it in layers. To redeem its appearance, a thousand trees were imported and planted and illuminated with mercury-vapor floodlights to vividly bring out their green.

The labor force assembled exceeded that for the IRT—24,500 at its height, with 600 engineers. Although the construction workers were unionized completely, there was an average of three strikes a day—

"largely," according to Hogan, "because we were using a lot of new materials, and the various trades were fighting for jurisdiction." Nevertheless, in the settlement of all contracts there was only one lawsuit, which the Fair Corporation won.

All the buildings were built to withstand wind pressures up to 100 pounds per square foot (actually encountered in the hurricane of 1938); and two of the foreign pavilions were erected by Parklap: the Belgian, which it built at a profit; and the Japanese, which it built at a loss.

One future employee of the firm, Priscilla Ogden Dalmas, monitored an architectural scale model of the fairground for the public's enchantment at a large table in the lobby of the Empire State Building. She kept daily track of the Fair's progress, and "whether a ditch has been dug for a new sewer," reported a journalist, "or a weather vane has topped a flagpole, Mrs. Dalmas gets wind of it."

Another future employee, however, had rather more to do with the Fair's success.

Although there were a number of impressive exhibits, including Futurama with the first limited access highway (by General Motors) and television (by A.T.&T.), when the Fair is remembered it is usually for its celebrated theme buildings, the Trylon and Perisphere. The Trylon, a needle-like triangular steel pyramid, and the Perisphere, a huge white hollow sphere, together formed the physical center of the exhibit and struck contemporaries as the exact symbols for its overall theme, "Building the World of Tomorrow." With their pristine geometry, gleaming white finish, and monumental scale (the Trylon was taller than the Washington Monument, the Perisphere 200 feet across), they represented both the structural principles of the column and the dome, and infinite aspiration beside the finite world.

Inside the Perisphere, visitors could stand on one of two revolving sidewalks to view "Democracity," a panoramic model and multimedia demonstration of urban life as projected for the year 2039. The lighting created a three-dimensional effect and "made you feel as if you were suspended in mid-air."

Both buildings were coated with a magnesite compound in stucco, and at night the illuminated Perisphere looked like "a huge iridescent soap bubble filled with moving clouds and mists."

The design of the Trylon, because of its tapering height, was controlled by wind forces rather than by dead and live loads; yet its balance was so successful that "when an observer stood with one foot on the Perisphere and the other on the ramp connecting it to the Trylon, only the slightest tremor could be detected in a 45-mile per hour wind."

The structural engineering for these extraordinary buildings was the work of Alfred Hedefine, a Waddell & Hardesty employee. He would join the firm in 1948 and figure prominently in its later history, but when asked when his tenure with the firm began he always dated it to this time.

The World's Fair opened on April 30, 1939, a day carefully chosen by President Roosevelt to coincide with the 150th Anniversary of George Washington's inauguration. It had the largest foreign participation of any fair in history (63 nations), and the largest attendance (44,932,978) in 354 days. Beyond that, its severely modern architecture, with its simplified lines and shapes (in contrast to the Columbian Exposition in 1893 which had fostered a neoclassical revival) did much to disengage contemporary American architecture from traditional styles.

The layout of roads at the Fair was next to flawless; and if General Motors' limited-access highway was a prime exhibit, Brinckerhoff's own fairground network was likely its match. Even then Brinckerhoff was successfully pioneering a new discipline that would have a profound and lasting effect on the development of America's roads. This was the so-called "traffic and earnings" report, which predicted revenue traffic for toll roads and supplied the basis for the bond issues by which they were financed. The first of these was for the Pennsylvania Turnpike, a project begun in 1927 and to which Brinckerhoff also contributed essential design.

Brinckerhoff laid out a four-lane highway, graded and aligned so that a loaded truck could travel at 40 miles per hour on the outer lane in each direction as traffic proceeded on the inner lane at 60 miles per hour. There were no grade crossings and at any given point there was a sight distance of at least 3000 feet.

The Depression delayed the project, but after the establishment of the Reconstruction Finance Corporation in 1932, a loan of $75 million was secured on the basis of Brinckerhoff's estimates and plan.

The traffic estimates were made over a wide area, and covered many divergent routes. But the final estimate had to be an educated guess, since "there was absolutely no precedent for a road of this type." Nevertheless, with regard to the first year of operation, Brinckerhoff hit the mark within 2 percent.

The Pennsylvania Turnpike was the first long trunk toll road in the world, and over the next three decades it led to a proliferation of turnpikes throughout the East and Middle West. The firm profited quite liberally from this, since Brinckerhoff and his protégés were the men to consult. In his traffic survey and layout of the road, he had established principles which would serve as models for all future roads of the type.

In early recognition of this fact, in 1941 Governor Thomas E. Dewey named Brinckerhoff to a commission to advise on the New York State Thruway, which had a projected length of 480 miles.

CHAPTER 12

I n a time of famine, it might appear the firm had a feast—a smorgasbord, in fact, of large projects spilling out of the professional cupboard and onto everyone's plate. And it is true enough that no one was starving, and that compared to others, Parsons Brinckerhoff had been doing well. On the other hand, remuneration was less than met the eye, for while the cost of a project itself might be substantial, the fee the firm collected after expenses was often comparatively small: $13,000 for the Buzzards Bay Bridge for example, $10,000 for the East River Drive, $2,500 for the Pennsylvania Avenue Bridge, and $8,500 for Hogan's services to the WPA. For Brinckerhoff's work on the Pennsylvania Turnpike, he received $23,500; for the firm's collective effort on the New York World's Fair—which had cost $126 million to build—$94,500. Moreover, some of these fees were spread out over several years. And, of course, some projects were done at a loss: for example, the Passaic River Bridge.

Douglas was looking anxiously ahead. Notwithstanding the volume of public work, it remained true that the private practice of engineering had fallen like so much watered stock with the Crash. By the late 1930s he had begun to search out new markets relatively unaffected by the turmoil of American finance.

In South America, where in reaction to the Depression there was a growing determination to develop new national industries free of foreign economic control, Douglas identified potentially ripening fields. In May 1938, he took opportunity by the forelock and opened an office in Caracas, Venezuela under Theodore Knappen.

The office got off to a propitious start, mapping out plans for a water supply system for Caracas, and drawing up plans for port development, irrigation, and housing, not only in Venezuela, but Ecuador, Colombia, and Peru. The design was done in New York under Thoresen (who at last in 1938 was admitted to the partnership*),

---

*Long barred, apparently, by "some rule affecting the foreign-born."

but occasionally Thoresen also went down to lend a hand. One day he went out into the field with a Colombian engineer to examine a site for a power plant and dam. He sat down on a hillside overlooking the valley, took a scrap of paper out of his pocket, and began to make a freehand sketch: this is the powerhouse, this is the dam, this is where the diversion tower would go. Then he produced a little scale, measured the distances, calculated the volumes of material required, and came up with a budget—which turned out to be just about right. "I don't suppose," remarked a former colleague recently, with appropriate awe, "there is one engineer in ten thousand who would attempt to do that. Then or today."

On another South American trip in 1939, he had a heart attack. His weight fell from 230 to 180 pounds, and he was prevailed upon to see a doctor for the first time, apparently, in his life.

The decade of the 1930s had in fact ground everyone down. Hogan, on the day the World's Fair opened, indulged himself, as he put it with wry restraint, "in a minor coronary thrombosis." Douglas, "worn out by his exertions," was dying.

In 1939 Douglas took steps to turn the management of the firm over to Hogan and to make him the majority partner. Brinckerhoff, also aging rapidly, was to remain the senior partner but with reduced participation. At the same time Thoresen with Hall and Macdonald (who had also just become partners) were to have their participation increased. Macdonald was to be in charge of the entire office force.

"This was the beginning," wrote Hogan, "of a dearth of partners and casualties to partners which persisted until the end of the Second World War."

At the end of 1940 (and a few months before he died) Douglas submitted his letter of resignation with this fine farewell:

It is with inexpressible regret that on account of illness I am forced to retire. . . . After 32 years of professional service with the firm I want to express to the partners and associates the great pleasure I have had in our joint work and association, and to express the hope and belief that our great firm will continue to work in the greatest harmony and friendship and will continue to give the highest grade of professional service even should the profit in a specific undertaking be negligible.

With your aggregate abilities nothing can stop you for a generation or more if you will pull together as the four original partners did.

The partners wrote back: "We do not know how we are going to get on without you."

CHAPTER 13

Upon the death of Douglas, Hogan became the iron man of the firm. Whatever had to be done he did, often at considerable physical cost to himself. During the period of preparation for the Second World War, and after its outbreak, what the firm was called upon to do was large, and unquestionably of paramount importance to the Allied cause. It is particularly to Hogan's credit that with so much war work in the offing he personally led the effort to establish standards for military construction contracts that would not admit of profiteering: namely, based on a cost plus fixed fee arrangement, as opposed to the fee as a percentage of the cost. The latter had prevailed during the First World War, and provided no incentive for keeping the cost down. Hogan triumphed; he overcame the opposition of the judge advocate general of the Army and the fixed fee arrangement was more or less adopted as the rule.

Hogan liked to recall that Parsons "insisted on two things: absolute integrity and public service," and it may be said of Hogan's own career that it sought to honor those mandates.

In 1940 President Roosevelt traded 50 World War I destroyers to the British government for air bases in Bermuda, Jamaica, Antigua, St. Lucia, Trinidad, and British Guiana, and made arrangements with the Cuban government for an air base at Camaguay. He secured large appropriations from Congress for their development, for the construction of new mobilization centers, and for the development of the Navy both at sea and on land.

About a month later Hogan was called into conference with the architectural firm of Voorhees Walker Foley & Smith by the Corps of Engineers, and in a joint venture arrangement negotiated a contract for the design of the airfields on Antigua, St. Lucia, Trinidad, and British Guiana. Somewhat later, Curaçao, Venezuela, and Dutch and French Guiana were added to the list. This was $100 million worth of work.

Hogan inspected the island sites and found that some of them had been selected without regard for common sense. The site at St. Lucia, for example, was on the extreme southern tip of the island, 40 miles from the nearest European settlement, and would have required five million yards of excavation to make a first-class landing field. The main site at Trinidad was 25 miles from Port-of-Spain, the principal city, "in the midst of a rain forest and jungle with an average recorded rainfall of over eighty inches a year."

Hogan concluded that the district engineer in charge was past his prime, and that he'd better put somebody competent on the scene. Hall was his first choice, but Hall had announced his intention to enlist if America entered the war. Thoresen was in charge of estimating and drafting and couldn't be spared. Brinckerhoff was not quite suitable for the task and in any case, as a member of the Chicago Subway Commission, was otherwise engaged. And Hogan "didn't thoroughly trust" Knappen. Macdonald, of course, like Brinckerhoff before him in 1917, was needed to carry on the regular business of the firm.

The man he finally resolved upon was Joe Wood, who had been in charge of the firm's work on the Fort Monmouth cantonment. Hogan canceled the cantonment contract, transferred its staff of 40 to Trinidad to start on the main base, with Wood as resident engineer, and reclaimed personnel from Bonneville and North Platte (both still in progress) to dispatch to the other islands. At the same time, he set up a central project office at 41 East 42nd Street under the architect Foley, with Quade as chief structural engineer.

But the site of the Trinidad base just wouldn't serve. Though intended as the terminal of all through flights from Africa, it was in the north center of the island at Fort Reed, drenched in fog and rain, and pilots complained of being unable to find the field. An emergency airstrip, which Wood constructed some distance away, was soon preferred. (Later, a more suitable terminus was secured from the Brazilian government at Natal, on the extreme eastern coast of Brazil, and Trinidad became a way station on the route to South America and Dakar, West Africa.)

Notwithstanding the difficulties, the landing fields at all four bases—Antigua, St. Lucia, Trinidad, and British Guiana—were completed before war was declared in December 1941, and over the next two years their squadrons helped drive German submarines from the Caribbean and the Gulf of Mexico, where at one point they were sinking so many ships as to endanger America's supplies of ore and oil.

Subsequent work on the bases at Curaçao and Dutch and French Guiana continued until 1943, when there was a change in the tides of war and the Germans were no longer a Caribbean threat.

Hogan had meant to direct all the Caribbean work himself (before relinquishing that role to Foley), but the military had promptly summoned him for an even more ambitious task.

In 1941 he was called into consultation with three rear admirals, Morrel, Harris, and Parsons (no relation to William Barclay), to discuss a new method of building drydocks that would cut their construction time by half. Harris and Parsons (both retired), with Morrel, the current chief of the Bureau of Yards and Docks, had devised their method for land-based docks, but were now designing floating drydocks on similar principles. The Navy, wanting to spread the work around, asked Harris and Parsons to collaborate with some of the country's larger engineering firms. Eventually four took part: Harris and Parsons, Parsons Brinckerhoff Klapp & Douglas, Moran Proctor & Meuser, and Fay Spofford & Thorndyke. Collectively, they were known as Drydock Engineers.

The docks they built, floating and in place, served capital ships, submarines, and cruisers, and at a cost of $200 million were "the largest contracts let by both the Army and Navy during the entire war."

The joint venture was organized in the office of Admiral Harris, with the firm contributing key personnel: Brinckerhoff as Hogan's alternate on the Board of Control; Bickel as head of the electrical-mechanical department (where he particularly distinguished himself); and Thoresen as head of estimates, where he served as a sort of court of last appeal. When the estimating department had given its opinion on some matter, Admiral Parsons used to say: "Now let's give it to Thoresen and find out how much it's really going to cost."

Some of the docks were enormous. The permanent docks for capital ships, for example, had interior measurements of 1100 by 110 feet, a depth of 50 feet, and a total mass of 250,000 yards of concrete. The equivalent floating dock—with a lifting capacity of 100,000 tons, enough for the biggest ship afloat—had a clear inside trough the size of three football fields laid end to end.

Permanent docks were built at the Brooklyn Navy Yard and in

Bayonne, Philadelphia, Long Beach, San Diego, Charleston, Boston, and Portsmouth, New Hampshire.* Work went forward night and day, and in January 1942 one of the Brooklyn docks floated a battleship. In the spring the Bayonne dock received one of the British Queens.

The floating docks were rather more remarkable. Built in sections for towing to advance bases, the manner of their construction made them almost impossible to sink. If a section was damaged, it could be unbolted from its neighbor and swiftly replaced. In fact, one part could be floated into the remainder and the dock could repair itself! Bickel explains how they worked:

A floating drydock consists essentially of a large pontoon, or a series of pontoons, with wide sidewalls. The pontoon is divided into watertight compartments which can be flooded to submerge the drydock to a depth sufficient for the draft of the ship to be docked. The wing walls, which are watertight, provide buoyancy to control the depth to which the dock is lowered. After the ship is floated in through one of the open ends of the dock, water is pumped from the ballast compartments and dock and ship are raised. Following repairs, the process is reversed and the ship sent on its way.

Their strategic importance was such that Admiral Nimitz included them in his index of secret weapons—though, as Bickel remarks, "a 100,000-ton dock would seem a hard thing to hide in a palm-fringed South Sea island bay." Nevertheless, they took the Japanese by surprise:

It was considered before this war that no fleet could carry on effective extended operations at a radius exceeding 2000 miles from a main supply and repair base. Compare this with the distances of 3700 miles to Japan, or 5500 miles to the Philippines from our only main Pacific base in Hawaii, and consider that many ships damaged in action during the early phases of the war had to travel all the way to east coast yards for repairs, to appreciate how urgent the need was for the establishment of fleet bases in advanced locations. . . . The development and successful use of mobile fleet bases by the navy was one of the main reasons for our success in the Pacific war, a factor which the enemy had apparently not been able to

---

*One for capital ships, begun at Pearl Harbor in the fall of 1939 as an early experiment, was completed before the Japanese attack. The Japanese, believing it couldn't possibly be finished, ignored it, and a few days later it had three destroyers under repair.

foresee. . . . Many a ship, too seriously wounded to survive the long trip home, reached the safety of one of these docks, to return again to fight and confound the enemy who had already consigned it to the bottom of the sea.

The Caribbean and drydock work was ultimately quite profitable, though in the interim a cash-flow crisis dangled the firm by a thread. Between the time an expense was incurred and money received there was a lag of several months, and as a partnership distributes its earnings annually and has no reserves, the interim borrowing to finance the projects was great. Unpaid fees and retained percentages (about $400,000 at one point) in theory provided ample security, but "an inexperienced bank vice-president," recalled Hogan, refused to authorize the requisite loans. "We tried ineffectually to explain the situation to him, and were obliged to transfer our account to another bank, which was glad to receive it."

Though there was a small margin of profit per piece on the drydock work, the volume was tremendous and the total profit was large. Moreover, notwithstanding the depletion of his staff, Macdonald in 1941 alone had somehow succeeded in doing over $100,000 worth of work.

The accounting department at this time consisted of one woman "who used to browbeat each and every poor underpaid draftsman to buy war bonds." Apparently, years of austerity had made her unable to adapt to new ways: when Macdonald, who could now afford it, decided one day to beef up her staff, she quit that night and was never seen again.

In 1943, in appropriate recognition of their role, Hogan and Macdonald replaced Klapp and Douglas in the name of the firm.

CHAPTER 14

D uring all this time Knappen's affiliated Caracas office had been thriving. And it might have continued to thrive but for a sudden scandal which, in after years, "officially made him a nonperson in the firm's history."

Knappen was a West Point colonel. He had the stance and the commanding air, great personal wealth, and a "romantic face" with a luxuriant year-round tan. Hogan never denied that he was a man of ability, but his "uncertain temperament" made the partners wary.

Around 1942, Knappen "became very insistent on a settlement of the Venezuela Company," said Hogan, which presumably meant he wanted the firm to buy him out. But the original agreement, drawn up by Douglas, was "obscure" because of certain clauses that allowed the company to avoid paying taxes and to treat its profits as capital gains. To postpone a settlement and the protracted negotiations it was sure to entail, Hogan offered to make Knappen a partner, despite his own misgivings and Macdonald's opinion that it would be a grave mistake.

As it happened, a timely opportunity arose in Brazil which the firm could scarcely decline. Brazil had obtained a $30-million loan from the Export-Import Bank for the development of an iron mine at Itabira, and for the reconstruction of a 230-mile railroad to connect it with the port of Vitória. Knappen had secured the contract, which stipulated that a partner be on the scene. Hogan (a little defensively, perhaps) recalled: "Even if he had not already been a partner, we probably would have made him a partner for this particular job."

However, prior to this Knappen had become romantically involved with a somewhat notorious demi-mondaine by the name of Betty Compton, a familiar figure at some of Society's wilder parties as "Gentleman" Jimmy Walker's wife. Walker and Compton were divorced, and soon after the Itabira contract was signed Compton married Knappen and together they headed for Brazil. However, when they reached Miami, she was told she could not accompany him further without a diplomatic passport, and in despair she took an overdose of sleeping pills. "On account of her former notoriety," said Hogan, "she made every front page in the country." Walker rather graciously intervened and secured the passport for her, but when Hogan flew down a few days later to see them in Rio he found them disconsolate in a hotel. Apparently, "on account of her unsavory reputation," the wives of various dignitaries had refused to attend a reception in their honor, and between outbursts of tears she announced that they could not remain.

Back in New York, Brinckerhoff was predictably mortified and demanded that Knappen be fired. Hogan, however, tried to salvage the situation and accompanied Knappen on an inspection tour:

> We went first to Vitória, the shipping point for the ore, then over 230 miles of badly laid out and badly maintained railway, to the iron mine at Itabira. The deposit was a veritable mountain of ore which was about 68% pure and one of the finest deposits in Brazil, if not in the entire world. About 250,000 tons of ore were being mined by hand annually, and shipped by rail to Vitória where it was loaded in tramp steamers for shipment to Europe. The locomotives on the railroad were wood-burning (since there was no coal in Brazil) and were on their last legs. The problem was to relocate and rehabilitate the railway, to reequip it with new locomotives and cars, and to equip the strip mine with modern machinery in order to raise the output to two million tons of ore a year.

Knappen and Hogan talked. Knappen said he had begun the surveys and would like to remain, but wasn't sure if he could. Hogan reminded him that he had solicited the job, had asked to be placed in charge, and that a design force had already been established in New York. Dejectedly, Knappen agreed to give it a try.

Hogan flew back to New York, but a week later received a call from the Bank in Rio that Knappen had disappeared. It was soon ascertained that he had withdrawn with his wife to a resort in the mountains.

> I discussed the situation with the partners and we all felt that we could no longer trust Knappen to represent us; that I had made a mistake in making him a partner and that the best thing the firm could do would be to get rid of him. I accordingly directed him to return to New York and received his resignation.

A year or so later Betty Compton died. Knappen's career, however, was not over. In 1945, after reaching a final settlement with the firm and apparently recovered from his romantic foray, he opened an office of his own in New York to which he attracted a number of quite capable people, particularly from Parsons Brinckerhoff. Within a remarkably short time it grew into a major firm. Knappen's strength was in foreign work, and he built upon it brilliantly. Today the company is known as TAMS (Tippets, Abbett, McCarthy & Stratton) and is comparable in size and diversity to Parsons Brinckerhoff itself.

Knappen's shoes in fact were not all that easy to fill, and at a time of a "dearth of partners" it took two to make up for his loss. Gerald T. McCarthy, his chief assistant, was made the special partner for Latin America and inherited the office in Caracas, while Samuel W. Marshall, who had been admitted to partnership for other work, replaced Knappen in Rio.

Indeed, the project in Brazil was all that one man could do. "It was like building a whole city," remembered Edward S. Sheiry, the resident engineer. "We built a mine, a railroad, a harbor, and a town." The town included schools, a church, and a jail, and a telephone system that took its current from 20 miles away. As the water supply system was in the next valley, a tunnel had to be bored through a mountain to bring the water in. However, during the rainy season work on the tunnel was hampered by mudslides. The ground, known as *canga* and composed of loosely agglomerated hematite particles, was baffling even to Sheiry, a specialist in soils. It had a hard crust on top but was soft underneath, "like stale gingerbread." Donkeys sank into it and had to be rescued with ropes; men often sank into it up to their hips.

Most of the workmen were impoverished locals, completely unacquainted with tunnel construction. When given molten lead for sealing the seams, they failed to caulk them tightly first with hemp, so the lead poured generously through into the pipes. Once hardened, it snagged dirt and debris, and the pipe backed up and was useless. The first time they laid it, it had to be ripped out.

Marshall did his best. Formerly chief engineer of the Pennsylvania Turnpike and an engineer "of all around ability," he had come into the firm in 1941 to help with the Caribbean air fields. But the partners, looking ahead, had really been counting on him to acquire and execute contracts for bridges and toll roads, which they correctly foresaw would be substantial after the war.

That mission for him would not come to pass. Aside from the sheer magnitude of the Itabira project itself, there were other problems with which he was helpless to deal. These gave him no firm footing, and like a workman in the *canga,* he sank. Hogan explains:

> The design, selection, purchase, and shipping of about 17 million dollars worth of machinery was handicapped by all the restrictions due to wartime priorities. . . . We were getting no cooperation from the Brazilian politicians, little progress was being made on the railroad, and nothing was being done in connection with the setting up of and care of the mining machinery. It was also apparent that the bank engineer was interested principally in the railway and cared little or nothing for the mine. This went on for about six months. . . . [By then] progress on the railroad relocation was so slow that the bank managers proposed to let a contract to an American contracting firm for finishing it up. He also informed me that there was a rumor that the Brazilian politicians intended to do nothing about the mine, and that they proposed to sell off some of the mining machinery for which they had a good offer. We accordingly wrote a strong protest to the Export-Import Bank through their engineer in regard to lack of cooperation from the Brazilians and lack of progress on the mine.
>
> Two days later Marshall and I were called into the American Embassy by the commercial attaché. Brazil had by this time entered the war on the side of the United States and was equipping a division to go to Europe. They had also permitted the United States to build a very large air base at Natal and were participating in the South Atlantic anti-submarine patrol. He stated that our protest might have the effect of throwing a monkey-wrench into the machinery and asked us to withdraw it.
>
> I stated that such questions were beyond our scope but that we had our reputation as engineers to maintain and refused to withdraw the protest. If, however, for diplomatic reasons they wished to disregard it, that was the prerogative of the Export-Import Bank and the State Department and not ours.
>
> In our opinion the whole thing smelled bad.

From the sound of his voice on the phone, Hogan realized that Marshall was breaking under the strain. Repeatedly, he urged him to come home. In November 1943 he did, and a few weeks later collapsed and died.

Hogan himself was tempting a similar fate. His overall health was deteriorating, and through a "deceased tooth" he had developed blood trouble which resulted in violent facial neuralgia or *tic douleureux.*

McCarthy, however, was confidently forging ahead. And he had just the right training for the job. A former expert on water supply and flood control for the Corps, in 1938 he had accompanied Knappen to Venezuela where together they had worked on the Orinoco River port of Ciudad Bolivar and on a flood control program for the Apure River. In 1944 he opened a branch office in Colombia, where he was responsible for five power plants—at Bucaramanga, Bogota, Manizales, Medellin, and Cali—and for the design of two similar plants in Argentina on the Atuel and Diamante rivers.

But McCarthy was not content. The reason had nothing to do with strains on his health; in a sense, it had to do with the health of the firm.

CHAPTER 15

W hen, in his letter of retirement, Douglas had urged his colleagues to "pull together, as the four original partners did," there was an implied warning in his words. In 1940, for all the frenetic industry of the firm, hard times were clearly not yet over; in the first six months of 1939, for example, the total net income of the firm from contracts was $40,460.22. After rent, insurance, taxes, payroll, phones, stationery and supplies, postage, and other miscellaneous expenses were met, the partners had only about $22,000 on which to draw. It was distributed thus: Brinckerhoff, $3900; Douglas, $4500; Hogan, $4500; Thoresen, $3000; Hall, $3000; and Macdonald, $3000. Even allowing for subsequent inflation, etc., this was hardly a Midas' hoard—especially for men who (outside of Macdonald) had long been in the vanguard of the firm. The war work, of course, soon turned things around. "In 1943," reported Hogan, "we were not only out of debt but flooded with money from retained percentages and realized profits." But along with success came personnel problems, for reasons Douglas may have foreseen.

The firm had a latent structural flaw, and prosperity shone a bright light on the crack. In a confidential memorandum to the partners dated May 6, 1943, Macdonald and Hogan suggested that some of the partners might perhaps be earning too much. A principal assistant, they pointed out, had just resigned because "with the large interest in profits maintained by the senior partners, he did not feel even a general partnership with a small interest" would earn him as much as he could make on his own. "This is the first time we have been told that a partnership in the firm did not look attractive, and it is time we sat up and took notice." The matter was urgent: in the economic boom expected after the war, the competition for talent would be keen. They went on: "Our young men represent a very considerable investment for us, because they are trained in our methods and acquainted with our clientele."

Gerald McCarthy, the special partner for Latin America, was among the restless. In an interim measure, the partners had met his terms. Now they tried to face the problem head-on:

> Those who are doing the work demand a bigger share in the returns. . . A partnership in a personal service organization does not represent any vested interests, and any attempt to treat it as such will only result in eventual decay . . . This means that the older partners, when they become less effective, must be ready either to retire or to give up a portion of their interest to reward those who are actually doing the work.

Obviously, the partners weren't quite pulling together; according to Hogan and Macdonald, they themselves brought in most of the work.

How much their efforts had changed the economy of the firm can be seen from the following figures: between 1936 and 1940, the average gross annual income was $182,842.30; the average net income before drawings was $58,602.61; and the average income of the partners from drawings and a distribution of remaining funds was $32,761.52. Between 1941 and 1945, on the other hand, the corresponding figures are $489,892.74, $172,101.23, and $140,014.68. In other words, in the second five-year or "war" period, the gross income had more than doubled, the net income had virtually tripled, while the actual income of the partners had more than quadrupled.

Hogan says in his *Memoir:* "I figure one year, that if I had received $10,000 more profit I could only have kept $1500 of it. It was evident . . . that some equitable means of greater distribution of the profits to deserving subordinates would cost the principal partners very little and would provide great incentives to subordinates who had done good work but were not of the partnership grade at the time. It would also eventually provide capable partners of which we were now in great need."

Timely as the memorandum was, the situation was rather more serious than they knew. For if most of the partners were no longer effective "business-getters," they at least had a reservoir of knowledge and experience on which their more active subordinates could draw. However, for reasons beyond anyone's control, six partners would soon be gone. The first of these after Marshall would be McCarthy, who in 1947 left to join Knappen, with whom he enjoyed an accelerated career.

PUBLIC WORK AND THE WAR: 1932–1945

*John P. Hogan, hydropower expert, began working with the firm in 1920. His exper-
tise led the firm to eminence in the hydroelectric field, and he became the senior
partner in 1939.*

The Bonneville Dam in Oregon, the first dam across the Columbia River. Hogan served as one of four consulting engineers.

In 1935 the firm supervised design and construction of a diversion dam, storage dams, canals, power conduits, a powerhouse, and transmission system for the Sutherland Project in North Platte, Nebraska.

The Trylon and Perisphere, theme structures of the 1939 World's Fair, were designed by Alfred Hedefine, who later joined the firm.

The steel skeleton of the Perisphere, in engineering drawings (right) and under construction (above).

As chief engineer for the 1939 New York World's Fair, Hogan (front center with scarf) supervised a large task force, drawing on the firm for his engineering core.

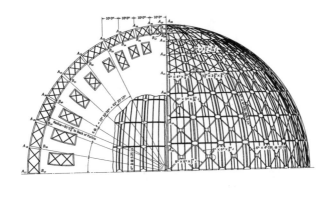

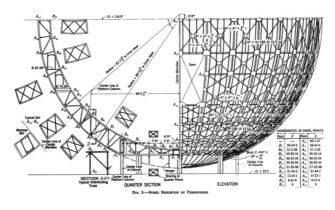

COORDINATES OF PANEL POINTS			
Point	f	Point	f
—	—	A_1	60-1-47
B_2	55-43-5	A_3	55-43-5
B_4	51-2-28	A_5	51-2-28
B_6	45-26-16	A_7	45-26-16
B_{20}	40-23-25	A_{21}	40-23-25
B_{22}	35-20-34	A_{23}	35-20-34
B_{24}	31-35-58	A_{25}	27-52-38
B_{12}	21-44-2	A_{13}	21-44-2
B_{14}	13-22-1	A_{15}	13-22-1
B_{16}	6-41-0	A_{17}	6-41-0
B_{17}	0	A_{17}	0

SECTION D-D
Typical Distributing Truss

QUARTER SECTION   ELEVATION

FIG. 8.—STEEL SKELETON OF PERISPHERE

The Buzzards Bay Bridge, with its lighthouse-like towers, was at the time the longest vertical lift span in the world.

In 1932 the firm's newly formed bridge department—Eugene Macdonald, John Bickel, Maurice Quade—designed a new Buzzards Bay Bridge to cross the Cape Cod Canal. It replaced a structure designed by Parsons when the canal first opened.

Eugene Macdonald (center) served as head of the firm's first bridge department. He and colleagues inspect the Jamestown Bridge in Rhode Island, 1939.

The St. Georges Bridge, a tied arch over the Chesapeake and Delaware Canal in Delaware.

Maurice Quade, for many years the bridge department's chief structural designer, with his trademark pipe.

A reservoir for Caracas, Venezuela. Planning a water supply system for Caracas in 1938 inaugurated the firm's work in South America.

A complex of maintenance shops for the Orinoco Mining Company's iron ore operation, Venezuela, 1939.

In Brazil the firm upgraded the iron mine at Itabira during World War II and rebuilt the railroad (above) to carry the ore to the port of Vitoria (below).

A weather station at the Cabra Corral dam site in the Juramento Valley, Argentina, 1945.

The British Guiana Airport, one of four air bases completed in the Caribbean before war was declared in 1941.

A drydock, designed for the war effort, Brooklyn Navy Yard, January 1942.

THE POST-WAR
ENGINEERING BOOM:
1945–1965

CHAPTER 16

I n late 1945 Hall returned from the war as a brigadier general. He brought with him William E. R. Covell, a major general and his commanding officer in the Far East, and also a substantial shipment of tariff-free PX whiskey which he stored (or floated) in the basement at Maiden Lane.

Both men had contributed briefly to the Caribbean venture in 1941 (Covell as an employee of the Corps); overseas, they built bridges and airfields in India and China, and helped construct through the jungles of Burma the famed Ledo Road. They were personally quite close, and under Hall's sponsorship Covell became a partner.

Though designated for foreign work (where new expertise was desperately needed), Covell's tenure with the firm proved brief. A tall, cheerful, breezy man with a great booming voice, he had graduated from West Point with "the class the stars fell on," so-called because a number, including Eisenhower and Bradley, became multi-starred generals. But despite obvious ability, his talents never blossomed, perhaps because his taste for the high life was a little too keen. When he took charge in Buenos Aires of an irrigation project for the Juramento Valley, one of his first requests was to have a new Cadillac sent down. And he once told Walter S. Douglas: "I'll do anything for you but work." Said Douglas ambiguously: "He was the best delegator of work I ever knew."

Other gaps in personnel were rapidly filled. Despite Hogan's and Macdonald's fears, most of the young men remained—among them, Rush F. Ziegenfelder, Lawrence W. Waterbury, and John E. Everson, as well as Bickel and Quade; others who had gone to war returned, such as Walter S. Douglas and William H. Bruce. Bruce would become the firm's chief field engineer; Waterbury, a transportation expert, appeared ready to fill Brinckerhoff's shoes; Bickel, upon Thoresen's retirement, became the firm's master of subaqueous tunnels; while Quade was well on his way to becoming "the country's best structural engineer." In Everson the firm had a budding project manager par excellence; in Ziegenfelder, an adept at highway and airport design; in Douglas, a sort of jack-of-all-trades with expertise in harbors and ports, pioneering ideas in mass transportation, and an aptitude for finance. Then in 1948, Alfred Hedefine joined, a bridge engineer of genius. And of course there were many new arrivals—William T. Dyckman, Seymour S. Greeenfield, Thomas R. Kuesel, and Winfield O. Salter to name four—who expertly filled out the ranks. In time, all would become partners, ensuring over the coming decades a bright succession of talent that in no way augured decline.

Meanwhile Hogan had finally retired—though under protest. His doctor had given him an ultimatum to step down by July 1946, but two pieces of unfinished business furnished a pretext for delay: the problematical negotiations for a final settlement with Knappen, which at last were coming to a head; and a contract for coal mine facilities in Turkey, on the southern coast of the Black Sea.* Then a third obligation arose, reminiscent of the World's Fair.

The Fair, in fact, had brought the job his way. Flushing Meadows had been chosen as the site for the United Nations, and as Hogan knew the terrain like the palm of his hand the mayor of New York quite sensibly appointed him to the Board of Design.

Hogan didn't expect his role would be much, since he had already done what he could for the site. But then John D. Rockefeller, in a fit of largesse, gave the city a substantial block of his property for the United Nations on Manhattan's East Side. It stretched from First Avenue to the East River, and from 42nd to 47th Street—and required the sort of complicated preparation that a subway usually entails. First Avenue, from 41st to 48th Street, had to be rerouted underground; three miles of adjacent streets closed off; 47th Street widened; a tunnel beneath 42nd Street removed; and two new approaches constructed for the East River Drive. Said Hogan: "What I entered into as a public duty developed into a large job with considerable prestige attached. There could be no thought of retirement until it was well under way."

He sped things along. First Avenue was the main problem: as it had never been projected for a subway, more utilities had been packed beneath it than under any other thoroughfare. To cut or disconnect

---

*Bickel prepared the proposal, but a Dutch concern "with more advantageous financing" got the job.

them would have incapacitated half the East side—although that had been allowed for in the plan. Hogan, however, secured an easement from the United Nations Corporation and reproduced all the utilities under it, so that with scarcely a break in service, the old lines could be dug up rapidly without having to be supported or rearranged. This common-sense bit of surgery, of great courtesy to the public, saved the city $5 million. And it meant, among other things, that the tunnel could be swiftly built as a cut-and-cover trench.

Having figured this out, Hogan retired on June 30, 1947. He bought a house in Santa Barbara to come home to, and two months later sailed for France, where for five recuperative years he basked in the Riviera sun.

Thus began a changing of the guard. Thoresen and Brinckerhoff were preparing to depart, Waterbury and Quade became partners, and Hall replaced Hogan in the name of the firm. But as partners, Waterbury and Quade belonged to a new class: they were limited partners, which meant that "they shared in the profits of the firm to a fixed extent. Anything beyond that went to Macdonald." This was in keeping with Macdonald's autocratic ways, "who referred to all his partners (except, of course, Hall) both in public and in private as 'my assistants.'"

Hall soon took over the business reins of the firm with Macdonald as senior partner, and in a significant move made Walter S. Douglas his apprentice.

Business was brisk. "Macdonald began hammering at us for budgets," remembers Everson, "trying to make businessmen out of his project engineers." Hall looked over everyone's shoulder and was "stern to get the right answer." He was *the* boss. Nobody crossed him."

He usually got what he wanted. "Right after the war," according to Douglas, "Venezuela owed us what seemed then like a lot of money, around $100,000. Hall went down with his rank and with General Covell—you know, they think a lot of generals and admirals down there—and the money came ripping in."

Hall's military connections proved useful in other respects. In 1947, through his friendship with the chief of the Corps of Engineers, the firm secured contracts for the expansion of two Air Force bases abroad. One was Holly Field (known later as Ernest Harmon) in Newfoundland, the other in Keflavik, Iceland. Holly Field was in reasonable shape, but the Keflavik base at the time was just a collection of quonset huts which America had acquired from Britain during the war because of its proximity to the sea lanes carrying American troops to the front. Through the construction of runways, taxiways, barracks, officers' quarters, roads, utilities, terminals, and so forth, both bases were upgraded for NATO use.

The work in Keflavik was touchy. The government in Iceland included a strong Communist Party, and all military personnel went about in civilian dress, with officers addressed as "Mr." rather than "Sir."

For a couple of years, Parsons Brinckerhoff was almost an adjunct of the Corps, which ran both projects from a special office established for the purpose at 130 Wall Street. Hall was in charge; Douglas and Bickel and later Greenfield were the project engineers. But a sizable proportion of the firm took part—more, in fact, than remained in Maiden Lane.

Hall didn't care to fly up for inspections too often. Though he had once told Bickel that in Burma "he would climb thousands of feet in a plane just to cool off," he seems actually to have had a fear of flying. On one trip to Iceland he fell asleep on an airport bench, and "it took three men to hoist him on to the plane." When they arrived in Keflavik, his first question was; "When is the next boat back?" It left two hours later and he caught it, never having seen the base.

On the other hand, the planes they had to use were somewhat unnerving: four-engine propeller-driven piston craft with old bucket seats that were wired together and shook like a mechanical mixer in a storm. Over the Baffin Straits, which looked like a solid sea of flowing ice, high winds would suddenly come off the Greenland icecap and toss the plane about like a ball.

Or hold it fast. Douglas remembers sitting in a plane off Greenland and staring down at some icebergs. After a time, he realized he was still staring at the same iceberg, in the same place, from the same angle: against the fierce headwinds, the plane had been unable to move.

Bickel was personally more drawn to the Nordic landscape, to the bluish mountain backdrop, the black volcanic rock exposed along the road. Reykjavik, the capital across the bay, reminded him of a fairy village, with its little red-roofed houses and the green waves pounding on the black rocky shore. In the summer, he remembers, "every afternoon you could see a rainstorm walking across the land," and in the winter the play of the northern lights resembled reddish-blue drapes "undulating hugely across the entire visible sky."

Two NATO bases in France, Dreux and Evreux, followed directly from this work, as did a contract to establish uniform design and construction criteria for all U.S. military airfields.

Then there was "Raven Rock." This was the first in a series of so-called "hardened underground defense facilities" supposed to be able to survive a nuclear strike. Also known as Fort Ritchie, and located on the Maryland-Pennsylvania border near Camp David, it was mined five hundred feet deep in the greenstone of the Catoctin Mountains, and equipped with living quarters, independent water, power, and air-conditioning systems, and vital blast protection devices to make it "self-contained." Thoresen developed the blast doors which sealed the access tunnels, and the firm worked closely with the National Bureau of Standards to develop a system that utilized the infinite mass of rock as a "heat sink." Though Raven Rock was designed as the joint Army-Navy-Air Force Communications Center, it was actually something of a trial run for the kind of construction it was. "It involved unprecedented concepts," said Douglas, "for which we prepared ourselves by visiting certain underground facilities in Sweden, Germany, and the Maginot Line. It was in Sweden that I gathered a knowlege of the use of rock roof bolts, and how to apply a thin coat of gunite to seal the joints so that they would always remain saturated with water. In addition, we pioneered in the installation of blast valves, blast doors, and underground utilities."

Under the spell of the Cold War, work went 'round the clock, 24 hours a day, 7 days a week.

Similar installations followed: the Strategic Air Command's Combat Operations Center, two regional headquarters for the Office of Civilian Defense Mobilization, the Army's "Sage" Combat Operations Center—and a culminating sixth, to be described in its place.

Meanwhile, in 1948 the firm decamped from Maiden Lane to 51 Broadway, a narrow five-story building between Broadway and Church streets which had once served as a Wells Fargo office. Despite a few vestigial amenities such as a marble fireplace and a Persian rug, it primarily resembled a "huge bowling alley," at least on the fourth or partners' floor. There was a long corridor, about 225 feet long, with

little cubby-hole offices off to the side and an equally small, improvised office for Hall at the end. "It looked great on paper," commented
a colleague, "but I don't think Hall used his scale."

Hall tried to put up bookcases from the firm's old library, but they
extended too far out into the room. So he bought some specially
designed furniture, as he said, "devoted to my personality," which
turned out to be diminutive antiques. One was an end table with
hollow books on top for containing bottles of bourbon.

"When you came to see him, it was awesome," remembers Everson.
"There was that long, long walk down the hallway and right at the
end of it was this great big man just staring at you and engulfing his
tiny desk. The desk looked like an extension of his stomach so that he
appeared to fill the room."

The building had its assets: for Macdonald, it was the fireplace; for
Hall, the Wells Fargo Truck dock in the basement where he planned
to keep his Army jeep. It was also not far from the Downtown Athletic
Club where Hall customarily went for a heavy lunch.

As the firm expanded, it took space in the neighboring buildings—
first 53 Broadway, then 55—and punched connecting passageways
through the walls. However, in some cases there was a discrepancy in
floor levels of as much as five feet, which were bridged with side
staircases. The place was like a "rabbit warren," remembers a client.
"Sometimes," said one engineer, "you had no idea where you were."

Before this lateral displacement, the highway and bridge departments,
under Waterbury and Quade, found lodgings at 75 Fulton Street. And
it could be argued that, for a time, the heart of the firm was there.

CHAPTER 17

The highway and bridge departments were closely allied, with the difference between them quite literally spanned by a separate highway bridge department.

The load the three were about to take up was immense.

Although the era of toll roads, begun with the Pennsylvania Turnpike of 1940, was delayed by the war, the automotive industry, through mobilization for the war effort, grew. Among other materiel, it produced 2.5 million military trucks, 6 million guns, 0.5 million aircraft engines, and 5 million bombs. The postwar surge in automobile production was therefore in part an adaptation to peacetime of a war industry wrought up to a fever pitch. Whether or not supply in this instance sparked demand, it is difficult to say, but the demand was consuming, and by 1955 production in the United States reached 7 to 8 million cars a year. The economic importance of the industry was such that it became the principal index of the nation's business health.

To accommodate this ferocious growth in traffic (which, both here and abroad, ultimately changed the very face of the globe), states scrambled to pave the way. Though before the war they had designed their own highways, the volume was now so great that they enlisted private engineers in legions to help. By 1954, 1500 miles of toll road had been laid in ten states, 1200 miles were under construction in 11, 3700 miles were authorized in 16, and 2600 miles were proposed in 9. A total of 28 states were active in various phases of toll-road development.

Roughly speaking, the highway work of the firm radiated from three centers—New York, New Jersey, and Virginia—and came under the respective jurisdictions of Ziegenfelder, Bruce, and Quade. If every project were to be named—particularly between 1955 and 1970, when the highway division accounted for a third of the firm's income—it might veritably mimic the biblical list of begats. But certainly the most prominent were the New Hampshire Turnpike, the Sagtikos Parkway, the Prospect Avenue Expressway, the Atlantic City Expressway, the Richmond-Petersburg Turnpike, the Bergen-Passaic Expressway, and New Jersey's magnificent Garden State Parkway, as well as sections of the New Jersey Turnpike, the New York State Thruway, and the Connecticut Turnpike. To these must be added such formidable "by-way" assignments as the complicated New Jersey approaches to the George Washington Bridge, the approaches to the Manhattan and Queensboro bridges, and numerous highway bridges of lesser or greater size—63 for the Atlantic City Expressway alone, for example; 17 for the New Hampshire Turnpike; 26 for the Rockland County section of the New York State Thruway; 35 for the New Jersey Turnpike; and 6 for the Pennsylvania Turnpike.

Overlapping projects of such number, weight, and measure do not admit of a tidy chronology. However, here and there a sort of pointer reading orients the map.

Everson recalls a gray season in 1950 when work on the New Jersey Turnpike had gone on for some months without pay. It was his first major job as a project manager, and he grew nervous as "anxiety began to build on the partners' floor." Everson had about 40 people under him, which was a large proportion of the office staff, and as he tells it: "One of the partners used to go around taking men off the project, so we wouldn't be wasting so much money." Everson would follow him, putting them back on. "We made a sort of a game of it, I suppose. But I really don't know why I wasn't fired."

One day he got a call from the Turnpike Authority in Trenton:

They said, "We have a check for you here for $130,000. Would you like to come down and get it? Or shall we put it in the mail?" I said, "This is a very appealing thing. It could really save my career. On the other hand, we're so busy on the job, I don't think I should take the time. Why don't you just put it in the mail?" So I went down to the partners' floor to give the good news to Quade. It so happened that General Hall was also in the office at the time. Hall said, "You really did not go down? Why didn't you send a messenger?" I said, "There weren't any messengers." "Look," he said, "for $130,000 I would gladly have been your messenger. Why didn't you send me? You know, if General Parsons could hear this he would roll over in his grave!"

The firm's crowning achievement in the highway field was the Garden State Parkway, probably the largest project Macdonald brought in and "the first high-speed toll road ever constructed in such a way that the traveler felt he was driving through a park." One hundred seventy-three miles long, and built over the course of 2-1/2 years (1952-55) at a cost of $330 million, it runs the full length of New Jersey from the New York State line in Bergen County down to Cape May—from the industrial areas of the north to the seaside resorts of the south. Along the way it incorporates major bridges over the Passaic, Bass, and Mullica rivers, and a 4,400-foot high-level bridge over the Raritan.

In its creation the firm had responsibilities of exceptional scope. Not since Parklap had it become involved in the fiscal management of a project in addition to design and construction; and at one time or another at least four principals took part—Macdonald, Quade, Hedefine, and Bruce—with a staff of 70 in Trenton alone, plus smaller field offices up and down the line. Moreover, 20 other architectural and engineering firms were placed under the firm's control.

The Parkway's distinguished success was followed some years later by the Richmond-Petersburg Turnpike in Virginia, where again the overall jurisdiction was comparatively broad.

Much of the New York State highway work came about as a result of the firm's special and lucrative relationship with the city of Albany. Since a mid-1940s redesign of the city's sewerage system and creation of a north-south vehicular bypass, Rush F. Ziegenfelder had practically acted as Albany's consulting engineer.

A small, quiet, rather gentle man who—untypical of partners—kept a drafting table in his office, he supervised the city's overall development, the design of adjacent highways such as the Schenectady Interchange-and-Thruway Spur, and especially the Albany Airport. Over the course of two decades he methodically transformed Albany's small municipal airfield into a major national hub. Across 800 acres of land he built runways, taxiways, a control tower, a terminal, air freight and crash-and-fire-rescue buildings, hangars, an aviation apron, a central bulk fuel storage area, and other facilities. In 1948, the airport handled 33,748 passengers; in 1969, one million; while the transport of mail increased fivefold and of other cargo, ten. The Greater Pittsburgh and Miami Airports likewise benefitted from his skills.

Remembered as a magnificent draftsman and designer, Ziegenfelder retired suddenly in the mid-1960s. Airport development thereafter continued under Dyckman.

CHAPTER 18

W hile the firm was excelling in largely new fields, it renewed its eminence—perhaps preeminence—in bridges and tunnels, the traditional mainstays of its name.

There was Florida's Sunshine Skyway, a majestic string of trestles and bridges across Lower Tampa Bay which featured prestressed concrete on an unprecedented scale. ("We did not pioneer in its use," said Quade, "but we stuck our necks out further than anyone else.") At Savannah, Georgia, the Eugene Talmadge Memorial Bridge—a high-level cantilever truss span with a suspended center supported by two cantilever arms—formed the dominant link in the nine-mile Savannah River Crossing. Off Staten Island, the firm surpassed itself with the Arthur Kill Bridge, which outranked the bridge at Buzzards Bay as the longest vertical lift-span in the world. Like its forbear built for a railroad, and normally raised for the passage of ships, the operation of its 558-foot, four million-pound span was "as smooth as a baby's skin." Hedefine, who collaborated on the bridge with Macdonald, compared it to "one of those brand new, high-speed precision elevators on Madison Avenue—the biggest elevator there is." In Texas, the firm's Pelican Island Causeway connected Galveston to 4,000 acres of undeveloped land by way of trestles, embankments, and a mid-channel, single-leaf rolling-bascule span. In Florida, the John E. Mathews Bridge, an 810-foot cantilever span across the St. John's River, was the state's first high-level bridge. Nearby, the Myrtle Avenue Bridge—a three-truss tied arch span—carried the Jacksonville Expressway over the Seaboard Coast Line rail yards. But perhaps the most outstanding bridge of the 1950s, and certainly the most original, was Quade's George P. Coleman Memorial Bridge at Yorktown, Virginia—the longest double-swing span in the world.

Before the war, according to a colleague, Quade had designed a high-level suspension bridge for the same crossing "to satisfy the Navy's clearance requirements, which were based on bringing a substantial part of the Atlantic Fleet into the coaling station upstream." However, the National Park Service and the Daughters of the American Revolution opposed it on the grounds that the towers "would intrude on the colonial atmosphere of nearby Yorktown Colonial Park" and disfigure the gentle landscape of fields and low rolling hills.

War stilled the controversy, and after reflecting on Pearl Harbor, the Navy decided it no longer wished to bring so many ships at once into a confined anchorage. The clearance requirements were considerably reduced and Quade took advantage of this to devise a unique double-swing span which kept the structure below the line of sight from the battlefield park, but permitted opening a center channel with unlimited height and sufficient width to pass the largest ships the Navy wished to bring in.

To further grace its architectural lines, Quade rested the shore ends of the swing spans on the cantilevered ends of the truss spans, instead of on piers.

This was all quite acceptable to the Daughters of the American Revolution, who also discovered to their delight that though the bridge was a huge mechanical contrivance, its operation could not have been more discreet. Though each of the swing spans weighed 1300 tons and was 500 feet long, at the touch of a button they swept "like a feather" without sound across the water, smooth as Hedefine's span at Arthur Kill.

Such piers as the bridge required were founded on underwater caissons—and these were extremely controversial at the time. "There is no glory in foundations, but there are many problems," observed Karl Terzaghi, the father of soil mechanics; and it was certainly true in this case.

Because of the poor bearing quality of the river bottom soil and the great depth of water at the site, Quade made the caissons hollow, both to reduce their weight and increase their stability with a wider base. His principles of design for the caisson plating, though common in shipbuilding, had never before been tried on a bridge, and most contractors were appalled. According to George Vaccaro, a specialist in foundations who worked closely on the bridge with Quade, five contractors bid for the job, and four of them stated "in no uncertain terms" that the caissons were "woefully underdesigned." Under the hydrostatic pressure, he was told, they would "fold up like an accordion." "I had many sleepless nights," Vaccaro remembers, for they pointed out that the bending stresses in the skin plate, when computed in accordance with the conventional method for a continuous plate on multiple supports, were considerably beyond the yield point of the steel. However, I maintained that this method of analysis erred very much on the safe side as shown by test results.

Nevertheless, I conveyed these misgivings to Quade.

Quade told us to proceed, but to include in the specifications that the contractor could base his bid on an alternate design. The lowest alternate bid exceeded the bid on the contract plans by more than $1,600,000.

I met with the low bidder on the original plans in the dining room of the Chamberlain Hotel in Fortress Monroe. I expected to hear the same objections. After a few minutes, I hesitantly asked, "What do you think of the caisson design?" He quickly replied (still munching on a succulent breakfast sausage), "I saw what you did and I agree with you."

From the depths, my spirits soared.

But that was not the end of it. To get the caissons to go down properly through 70 feet of soil, they were equipped with a series of pipe openings around the tops of their perimeters and projecting through the cutting edge at the bottom. Pressure water jets through the openings loosened the soil allowing the caissons to sink. That worked well enough until they hit an unexpectedly hard layer of marl. Roger Stevenson, the resident engineer, consulted with the contractor and came up with a solution. He called Quade. "Instead of a water jet in one of the pipes," Stevenson suggested, "I could drop a thin tube of dynamite through it and blow out the bottom." There was a long silence on the other end of the line. At length Quade said, "If you think that's the right thing to do, why don't you try it?" The caisson went down beautifully.

As bridges above, so tunnels below—or through. In 1954 there was the difficult Lehigh Tunnel through Blue Mountain on the northeastern extension of the Pennsylvania Turnpike, which required exploratory horizontal borings of record length, and provided the firm with its first opportunity to design the ventilation and lighting systems for a vehicular tunnel. With this project Sanford Apt and Seymour S. Greenfield demonstrated an in-house capability in the mechanical/electrical field that had previously been sought on the outside. Other projects of this period were the double-bore West Rock Tunnel on the Wilbur Cross Parkway near New Haven, the Baytown Tunnel under the Houston ship channel in Texas, linking the industrial cities of Baytown and LaPorte; and the first of two tunnels under the Elizabeth River between Norfolk and Portsmouth, Virginia. It was in the course of building the Baytown Tunnel by sunken tube that Thoresen, on the eve of his retirement, passed on his mantle to Bickel.* The Elizabeth River Tunnel fully proved that the mantle fit. Soon thereafter Bickel collaborated with Quade on a project that Quade would regard as the crown of his career.

This was the first Hampton Roads Bridge-Tunnel, which went over and under the ship channel between Norfolk and Hampton.

As the Virginia harbor on Chesapeake Bay, Hampton Roads is the main channel to the most important naval base on the Atlantic coast. A crossing by bridge alone was therefore out of the question, as interfering with naval shipping, while a vehicular tunnel three miles long would have been almost impossible to ventilate. In a completely novel solution, Quade and Bickel combined the two: a 1.5-mile long tunnel in midchannel that emerged on man-made portal islands connected by concrete trestles to the opposite shores. Eighteen miles of approach highway, including 20 bridges and one substantial bridge over Hampton Creek, completed the link between Norfolk and Hampton.

Tube sections of the tunnel were fabricated in Chester, Pennsylvania, towed through the Chesapeake & Delaware Canal and down Chesapeake Bay to an outfitting yard at Lambert's Point in Norfolk. The artificial islands, each about 1500 feet long and 250 feet wide, were constructed by hydraulic fill on the edge of shallow shoals. Though the north island lay in protected water, the exposed south had to be defended from storm wave erosion by a massive belt of heavy rip-rap stones— concrete monoliths poured in place, and granite blocks that weighed over ten tons. Concurrently, the contractors had to become "expert in the sex life of oysters" so as not to disturb the valuable beds which flourished in the silt nearby.

Though the length of the tunnel had been reduced by half, its ventilation requirements led to the introduction of vane axial fans (instead of the conventional centrifugal type) with a pitch automati-

---

*It also gave Thoresen an amusing opportunity to poach a bit on Brinckerhoff's terrain. Asked to estimate the traffic for the tunnel, Thoresen contacted a local Boy Scout master who stationed his troops along the way to make a count.

cally adjusted to the volume of traffic, and of such tremendous power that the ceiling had to be upheld by stiff suspenders to resist the suction updraft of the exhaust. One day when a painter was working in a duct, someone accidentally turned on the fan. The wind tore the pail from his hands, and in an instant pinned him like a butterfly against the wall. Screaming wildly, he clung to a ceiling strut so as not to be sucked into the turning vanes.

§

While bridges, tunnels, highways, airports, and military installations monopolized most of the staff, other disciplines were not permitted to lapse. Thus a random sampling of the additional work during this period includes: a new pier for the Hoboken Marine Terminal, the widest pier in the New York-New Jersey port; stability studies for offshore drilling platforms; a marginal wharf at 39th Street in New York City built over two tubes of the Lincoln Tunnel; an analysis of the structural frame of the Roosevelt Raceway grandstands; a sugar wharf for the American Sugar Refinery near New Orleans; a comprehensive study to establish a practical program of flood insurance rates; and the mechanical and electrical features for a new Bloomingdale's department store in Hackensack.

Some opportunities, like the coal port in Turkey, inevitably slipped out of reach, such as a shield-driven tunnel under the Riachuelo River in Buenos Aires and a sunken tube tunnel for Porto Alegre in Brazil. Others popped up in odd or unexpected ways. One night in 1948, for example, Macdonald was awakened by a phone call telling him a New Jersey pier had just collapsed into the river with a million dollars worth of Campbell's Soup. With that minor toxic spill a long and remunerative relationship with the New Jersey Port Commission began.

Another waterfront calamity was narrowly averted because the engineer in charge, George Vaccaro, was morally alert.

Vaccaro had been carefully designing a drydock for a company in Camden, New Jersey. However, the company was so eager to start that, he recalls, "I wasn't sure which was going to get there first, the shovel or the plans." Subsequently, he was told his plans would cost too much, and an executive of the client company unscrupulously pressured him to doctor his report to the company board. Vaccaro refused. The executive told him "You're designing the drydock like the Romans did!" Vaccaro replied: "Yes, and before the Romans like the Greeks did, and before the Greeks like the Syrians did, because gravity and buoyancy and Newton's third law of motion were there then just as they are today, and these you have to respect!"

§

Perhaps no engineer in the firm has had quite so much to do with the actual construction of projects as William H. Bruce, a gregarious, abrupt and open man who in 1950 was appointed chief field engineer.

Born in England in 1908, Bruce began his career with the county engineer at Barnstable, Massachusetts. In the early 1930s he worked for the firm on the Buzzards Bay Bridge, then returned in 1939 to work on the Jamestown and St. George's bridges. During the Second World War, he served with the Corps of Engineers on the construction of the Cape May Canal and various harbor defenses, and then as a naval officer with the Seabees in the Pacific where he helped assemble the floating drydocks that Bickel and Thoresen designed.

But it was as chief field engineer that he established his reputation. A man with "a way with contractors," he could "get men to do things," said Macdonald, "that they thought it was not in their contract to do." An official of the New Jersey Highway Department once called him a "Cassius," then held up the contract they negotiated as a model for the state. He also had a gift for putting his staff at ease. On the first Elizabeth River Tunnel, one engineer remembers, "Bruce came down to the job from the New York office, and he was God Almighty out of New York. He said, 'O.K. boys, when we finish work, I'll buy you a beer at the bar.' "

Bruce was especially close to Macdonald and Quade. "Most of what I did before I became a partner was under their direction. Macdonald was a very astute businessman, Quade a most capable and conscientious engineer. From these two I formed my idea of what a consulting engineer should be."

In the early 1950s, with the sudden large increase in projects requiring field supervision and upwards of 200 people in the field, he had to perform like a virtuoso juggler. "In addition to the Baytown Tunnel," he says, "at one time I had to staff and oversee our efforts on the Sunshine Skyway, the Mathews Bridge, the Yorktown Bridge, the Talmadge Bridge, a section of the New York Thruway, a section of the

New Jersey Turnpike, and our classified work at Blue Ridge Summit [Raven Rock], where I also served for a time as resident engineer. My duty was to give the resident engineers whatever assistance and backing I could—men like Peter Hackman, Stanley Johnson, Joseph Whiteman, Joe Goldbloom, Bill Pease, Harold Wombacker, Bert Tryon, Roger Stevenson, and others who formed the nucleus of the field corps."

Especially after 1952, when Macdonald named Bruce project manager for the Garden State Parkway, he found himself unable to monitor things as he wished. Accordingly, at his initiative, the firm established a central Construction Department under his direction in the New York Office to bring more efficient administration to field activities and to keep the partners themselves more abreast of their own jobs.

Over the years, the Department also became an indispensable source of reference for the New York office, as a rich repository of field reconnaissance and construction "know-how."

The formation of the Construction Department was a major break with the partnership tradition, for it centralized certain jurisdictional functions that the individual partners had always jealously guarded as their own. As such, it foreshadowed things to come.

In 1954 Macdonald made Bruce a partner, along with Bickel and Ziegenfelder.

Two years later Macdonald retired, presumably content. He had accomplished what he had set out to do, to rebuild the firm, and under his tenure it experienced its greatest relative growth, from 150 to 500 employees. Moreover, every one of the remaining partners was new.

"He was a real Yankee," said Douglas (momentarily forgetting he was a Scot), "and an aggressive businessman. He could out-trade a horse trader." Macdonald said of himself: "In the pursuit of new clients, I was the best sitter in the outer office of any man."

CHAPTER 19

In the history of the firm up through Macdonald's time, there is no evidence of a struggle for the succession. Perhaps it did exist; it may even be likely; but there is no evidence for it, at least in what is preserved. One might wonder, for example, how Hall and Macdonald got along. Hall had been with the firm since 1915, Macdonald since 1932, yet when Hall returned from the war Macdonald was Hogan's anointed. On the other hand, when Hogan retired and the name of the firm was changed, Hall's name went into Hogan's slot, leaving Macdonald's at the end. Parsons Brinckerhoff Hogan & Macdonald became Parsons Brinckerhoff Hall & Macdonald—an extremely irregular rechristening in the history of company names. Obviously, this was an accommodation, as was the office that was improvised for Hall at 51 Broadway. "They couldn't find a space for him," recalls Everson, "but thought he should be next to Macdonald because of his position." (Macdonald's office, with the marble fireplace, was at the end of the corridor off to the side.) In any case, equal or not, after Hall died suddenly of a heart attack in 1951, Macdonald was the undisputed king.

Now Macdonald, in planning the succession, did something quite interesting—and he did it twice. In 1947 he made Waterbury and Quade partners on the same day, and in 1952 he made Douglas and Hedefine partners within the space of a month. In both cases, he paired a man whose primary aspirations lay in the business of the profession with a completely professional engineer.

The strategy is obvious; but it failed. Within each pair there was a falling out. To begin with, Waterbury claimed at once that Macdonald had laid hands on him first. So when in 1948 a new brass partnership plaque was engraved (as was periodically done) with the names arranged in descending order of seniority, Waterbury expected his name to come before Quade's. For reasons of symmetry, however (the innocence of art!) their order was reversed—and Waterbury promptly had the plaque destroyed.

In retrospect, his cause was probably lost. Whatever his technical priority, he had no practical constituency. He was a lone wolf in a specialized field.

That field was the discipline of revenue estimates that Brinckerhoff had pioneered. Waterbury developed it with zeal. By the late 1940s and early 1950s he was preparing surveys and reports for garages, parking lots, and toll bridges in New York, Indiana, Florida, Illinois, and Missouri; and for highways and turnpikes in Pennsylvania, Ohio, New Hampshire, Texas, Florida, New Jersey, Delaware, and Wisconsin.

But in a sense he became too energetic at what he did. His estimates outraced the firm's quest for highway design (a far more lucrative activity), and as the same firm—on the presumption of a conflict of interest—was seldom allowed to handle both, he estimated the firm right out of some coveted jobs. "We didn't design the garages," says Bruce, "we didn't design the lots; we didn't design one damned linear foot of the Illinois or Ohio turnpikes." Moreover, some of Waterbury's revenue predictions were overly optimistic.

Nevertheless, Waterbury was highly regarded, successful in his own way, and when Macdonald prepared to retire in 1956, there was some speculation, at least among the troops, as to who would take the reins.

Personal problems, however, did him in. It was tactfully announced in the local papers that he was leaving to form an office of his own, whereupon the partners bought him out and took his portrait off the wall.

Quade took control, but now with Douglas (successively tutored by Hall and Macdonald) in charge of things at the financial end. In a sense, this put the succession planning further out of joint. Douglas moved into Hall's little office, and the name of the firm was changed to reflect the new regime. But at Parsons Brinckerhoff Quade & Douglas, Quade was the captain of the ship.

By all accounts, Quade was a man conservative in every way who worked very hard and expected as much from his staff ("the more you turned out," said one, "the more he gave you"). Macdonald called him "a pillar of rectitude and integrity" and "the best structural engineer in the United States."

He had a broad, square, clean-shaven face with the pouchlike cheeks of a chipmunk and habits of frugality to match. It is said that "he demanded an accounting to the penny," and it would appear this was literally true. When one engineer submitted a bill for his work on the West Rock Tunnel, Quade replied:

> I have received your two expense accounts reporting on the $300 which has been advanced to you for your expenses in connection with this project. In general, your expenses seem to be unusually large, especially in connection with a field job. . . .
>
> We must disallow the item of $14.08 for work clothes. It has never been our practice to purchase ordinary personal clothing for any of our men, either for field or office work. I notice also that you have consistently reported your expenses for meals as an even number of dollars, varying from $4.00 per day to $7.00 per day, and averaging $6.00 per day. In general, this might be apportioned on the basis of $1.00 for breakfast, $2.00 for lunch, and $3.00 for dinner, which is a considerably higher standard of living than that maintained by the partners, even when, of necessity, we stay at the best hotels in the larger cities. I am reminded of the fact that during our last conference in New Haven on September 30th, I paid the bill for lunch at the Cape Cod Restaurant for eight people, including the deputy highway commissioner, and the total amount, including the tip, was under $7.00.

This was typical. Arthur Jenny, an outstanding engineer who worked on the Walnut Street Bridge in Harrisburg, Pennsylvania, remembers that Quade dispatched him to Harrisburg for inspection trips at night so he wouldn't lose a day's work at the office. And while in Harrisburg, Quade "made sure I stayed in the cheapest hotels and ate at the cheapest diners."

Not surprisingly, Quade was also a stickler about dress. One evening he dropped into the Reports Department and saw two stenographers leaving in slacks. He said to the department head: " 'Do those girls work for us?' I said, 'Yes, of course.' 'Well,' he said, 'what are they doing in slacks?' 'Well, Mr. Quade,' I said, 'they are going bowling tonight, so they went to the ladies' room and changed. But they don't wear them during office hours.' 'Oh, good,' he said."

Nevertheless, it was Quade who put an end at the firm to the five-and-a-half day week (because he was an avid sports fan, and liked to go to Saturday games). And as reserved as he was, he made it clear that if anyone had a professional problem, he could knock on his door. "At any hour of the day or night," says Bruce, "at home or in the office, he was ready to review problems we had in the field." "I could always go in and discuss things," recalls Everson, "and would always come out with a well-thought-out answer. Sometimes he even made me believe it was my idea." Milton Shedd remembers that one Friday afternoon "we discussed how to frame a barrier gate to a concrete trestle. On Monday Quade came in and handed me two pages of detailed calculations and sketches. It was a very unusual thing for a partner to do."

One of Quade's great virtues, apparently, was poise in a crisis. "If there had been some trouble on a job because of an oversight," says a colleague, "he didn't waste time on a reprimand. He concentrated on the solution." Solutions were usually quick to come. According to Thomas Kuesel, "Quade had the most incisive mind of any engineer I ever met. On many a Monday morning, he would come into the office, spend an hour reviewing the weekend football games, and suddenly switch to a discussion of some technical problem that a group of us would be struggling with. His first question would clearly define the central issue, and with one or two more questions and a puff on his pipe, he would have indicated how to proceed and where to look for the proper answer."

Quade's pipe was his trademark: it was always in his mouth, and he "puffed it like a steamboat," occasionally scattering coals in his wake.

It might have been the emblem of the firm, which was steaming ahead. Much of what Macdonald began Quade safely saw through, and even a partial enumeration of the projects during his tenure reads (to Macdonald's begats) like the Homeric catalogue of ships: offshore drilling rigs for the continental shelf in the gulf of Mexico; a sewage treatment plant at the confluence of the Susquehanna and Chenango rivers for the city of Binghamton, New York; the Charlotteburg concrete gravity dam for the water supply of Newark, New Jersey; a sewerage system for Baton Rouge, Louisiana, and its environs; the Shenango Dam and Reservoir in Pennsylvania; and a river intake and four-mile-long water supply tunnel for Wayne County, Michigan.

Many of the water supply projects were directed by G. Gale Dixon, who had been with the firm since the Caribbean venture in 1941. Difficult to get along with, gruff and crusty, he never became a partner, though he made sure he was listed as an equal on any project proposal that touched his domain.* In that domain he was Hogan's true successor, and after he died in 1959 the firm never quite made up for his loss.

Bruce, however, did his best to fill the gap. He negotiated a sewer master plan for the City and Parish of Baton Rouge, Louisiana, which eventually led to design and construction management for some 80 miles of trunk sewers with pumping stations and three treatment plants. There were also sewerage studies and the development of a flood relief program for Hamilton, New Jersey. In these endeavors Bruce, of course, had expert help, and among those on whom he could always rely were Eugene Hardin, Herbert Kauffman, Alden Foster, and Evan Vaughan.

During this time, the firm's waterfront engineering was also particularly extensive, reflecting one of Douglas' chief interests and the expertise of a new partner, Admiral Robert H. Meade. The firm was retained for freight and passenger terminals at Port of Spain, Trinidad; bulk cargo facilities at the Portsmouth Marine Terminal in Virginia; overall development of the port of Toledo, Ohio, the largest transfer facility for coal and ore in the country; a yacht harbor in the Bahamas; a railroad coal terminal for New York harbor; the modernization of three major ports in Colombia, South America; a deep water harbor terminal on the island of Antigua; installation of Edward Sheiry's patented marine hose derricks at numerous refineries, including one in south Wales; a long-range modernization program for the port of Callao, Peru; the Brooklyn Marine Transfer Station; a redevelopment plan for the port of New Orleans; a two-pier passenger and cargo terminal in Manhattan; two new piers for the Hoboken Marine Terminal; and a study for the redevelopment of thirteen piers on Staten Island.

In the midst of this overall boom, even the highway work gathered speed. With the completion of the Garden State Parkway in 1956, the firm was in a particularly strong position to take advantage of new opportunities in New Jersey and elsewhere that were stimulated by passage of the Interstate Highway Program in 1954.

Douglas Sayers, a toughened veteran of the Pennsylvania Depart-

---

*The title block read: "Parsons Brinckerhoff Hall & Macdonald and G. Gale Dixon."

ment of Transportation and a stickler for good, plain, basic design, headed the highway department with support from, among others, Winfield Salter, Martin Rubin, Richard Duttenhoeffer, Charles Louis, Mel Kohn, Richard Shellmer, and Arthur Jenny. Their talents were complemented in the highway bridge department, by those of Michael Fiore, George Vaccaro, Milton Shedd, and William Dyckman.

When Sayers became ill, Duttenhoeffer, who had helped design the highways for the first Hampton Roads project, assumed charge, eventually replacing Sayers as department head. From 1949 to the mid-1960s Duttenhoeffer guided and sustained the firm's effort in highway work, even as the firm turned increasingly toward rapid transit.

Meanwhile, Parsons Brinckerhoff had established branch offices in four cities, field offices in nine, was billing about $5 million a year, and had moved out of its bowling alley quarters to 165 Broadway, a skyscraper owned by U.S. Steel with a lobby expensively finished in Italian marble and, more quaintly, with the last large-scale hydraulic elevators in New York.

In June 1966 Quade died, a few days before his scheduled retirement. In his tenure, he had not been as aggressive as Macdonald, perhaps, but he had followed certain sensible principles which could scarcely lead him astray. "Successful management," he wrote, "requires men who can act as professional men, who can work with their equals as well as their superiors and subordinates, who can be cooperative at all levels, and who basically have the best interest of the client, the firm, and the profession at heart." Such management, he added, is likely to prevail so long as "control always remains firmly in the hands of professional engineers."

THE POST-WAR ENGINEERING BOOM:
1945–1965

Upon Hogan's retirement in 1946 Eugene Macdonald (above) emerged as the senior partner; Gene Hall (below) took over the business reins.

A work crew at Raven Rock, also known as Fort Ritchie, the first hardened under-ground defense facility.

The switchboard at 51 Broadway where the firm moved in 1948.

The Lehigh Tunnel on the Pennsylvania Turnpike, which utilized modern ventilation and lighting design.

Ernest Harmon Air Force Base in Newfoundland, one of several air bases for which the firm provided design and construction supervision.

Construction of the Garden State Parkway was most difficult in the urban north.

The Garden State Parkway—the firm's crowning achievement in the highway field. The 173-mile long toll road, completed in 1956, connects New Jersey's urbanized north with its rural south.

Toll plaza on the Richmond-Petersburg Turnpike in Virginia, another major highway project, begun in 1955.

Work on the intricate New Jersey approaches to the George Washington Bridge.

The twin bridges in Woodbridge, New Jersey. In all, the Garden State Parkway included 459 bridges. The one on the right is one of the largest.

Maurice Quade, who succeeded Macdonald in 1956 as head of the firm.

Partners in the early Quade era (1956) were (from left to right): William H. Bruce, Jr., John Bickel, Rush F. Ziegenfelder, Maurice N. Quade, Walter S. Douglas, and Alfred Hedefine.

Florida's Sunshine Skyway, completed in 1956, featured use of prestressed concrete on an unprecedented scale.

The George P. Coleman Memorial Bridge, the longest double swing span bridge in the world, was Quade's supreme accomplishment in the field.

The Baytown Tunnel under the Houston Ship Channel in Texas, one of several subaqueous tunnels designed between 1945 and 1965.

John Bickel and Michael Fiore, structural engineers, on a vertical lift bridge project on the Ohio River, Louisville, Kentucky, 1957.

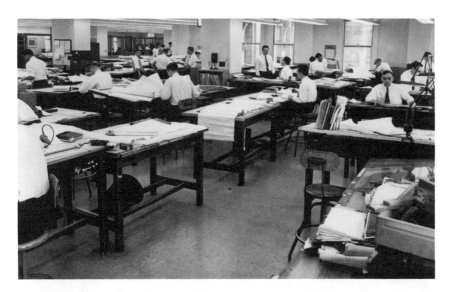

By 1957, when the firm moved to 165 Broadway, the bridge department had grown from three engineers in 1932 to over 30.

The Port of Toledo, Ohio. Work begun in 1957 continued for ten years on this 125-acre, 18-berth complex.

The Hampton Roads Bridge Tunnel in Virginia—a 3.5 mile long combination of trestles and tunnels. Uniquely, the tunnels emerged on man-made islands. The first bridge tunnel (on the left) was completed in 1957, the second in 1976.

The firm's partners in 1965 (from left to right): Alfred Hedefine, William H. Bruce, Jr., Maurice N. Quade, Walter S. Douglas, Rush F. Ziegenfelder; (standing) Robert H. Meade, Seymour S. Greenfield, John E. Everson.

TO KNOW ONE'S OWN ESTATE:
1965–1985

CHAPTER 20

P arsons told Walter S. Douglas as a boy: "If you ever grow up to be the man your father is, you'll be grateful." Douglas succeeded Quade in 1965, and he was very much his father's son: a capable engineer, a shrewd and farsighted businessman, occasionally witty, generally well-spoken, and absentminded beyond belief. Almost every day he managed to leave his hat or raincoat on the New Jersey ferry (from which his secretary would dutifully retrieve them); when he traveled abroad, "he left his papers and belongings all over the world." Sometimes he would forget to tell his wife that he was traveling, and she would call his secretary in the middle of the night and ask, "Do you know where my husband is?" One evening a colleague showed up at his door as dinner was being laid and said, "Are you ready?" His wife said, "Ready for what?" And Douglas replied, "Oh, didn't I tell you? I'm catching the night train for Atlanta in an hour." While working on a classified government project, he had to have top-secret documents strapped to his arm.

This is simply the way he was. "When he concentrated on a problem," according to Everson, "the outside world meant nothing."

Born in New Jersey in 1912, Douglas accompanied his father to Panama in 1918, where he remembers being poled down the Chagres River past "massive growths of orchids, a great profusion of birds, and many alligators." On Armistice Day his father, as Acting Governor of the Zone, declared a school holiday, and "I have never since," said Douglas, "been such a hero as I was to my fellow schoolboys in Panama."

He set his sights on Lehigh College, his father's alma mater, but his father said, "I don't think the girls in that town are any good for you," and sent him to Dartmouth instead. There, according to a classmate, he had "some pretty fair orgies" anyway, and graduated in 1933. In 1935 he earned an M.S. from Harvard.

Douglas took his first job in a steel shop in Nashville, Tennessee, where he worked ten hours a day, half a day Saturday, and made $18 a week — which wasn't bad: "I spent $7 for a very nice room and board, run by a lovely landlady who made up my lunches and did my laundry. To that was added $3 for carfare, cigarettes, care of my clothes and miscellaneous. I put $5 in a savings bank each week. The rest I blew on Saturday night and, believe me, I was the best-heeled bachelor in the state of Tennessee."

In 1937 Hogan made him his assistant on the World's Fair, and from 1939 to 1942 he was apprenticed to the firm as a structural engineer. When the war came, he was sent first to Fort Monmouth, New Jersey, to help construct a new headquarters for the Signal Corps, and then to a base in Trinidad in 1941. "It was the only job I ever worked on I was ashamed of. Aside from the engineers, everyone was scrambling for glory." From there he was called to active duty as an officer in the Navy Civil Engineering Corps, assigned to the Seabees, and by Thanksgiving 1942 was in New Caledonia building piers to help break the bottleneck in supplies for the Guadalcanal campaign. After amphibious training, his battalion landed with the Marines at Rendova. Later, at Munda, he kept a road open through the mud so the army could advance. "I stood beside that road for 48 hours," he recalled. "The battle was just over the brow of the hill. A colonel came by and gave me a Japanese rifle and said it was something to remember them by."

Douglas rejoined the firm in 1946 and took over the South American desk which he found in shambles. By 1947 he was Bickel's assistant on the Ernest Harmon Air Force Base in Newfoundland and the Keflavik base in Iceland, and when Bickel went to work on the Baytown Tunnel, Douglas became project manager for both.

It was by his work on Fort Ritchie, or Raven Rock, however, that Douglas really made his mark. "We made a real success of it," he says, "both professionally and financially, which is usually the case — either you come out well in both or well in neither. As a result of Fort Ritchie, when Hall died, the remaining partners — Macdonald, Quade, and Waterbury — promptly made me a partner. Within a few weeks, Macdonald told me that in addition to my engineering assignments he wanted me to take over the administration and financial affairs of the firm."

Fort Ritchie led to four other hardened defense facilities, "each an improvement on the last," says Greenfield who served as project manager for them all, "in response to new generations of weaponry." Then came the sixth — "the Dr. Strangelove thing up in Colorado

Springs," as one former partner put it, "that would sustain a near direct hit of an atomic bomb, and all the generals would be there": NORAD.

NORAD, or the North American Air Defense Command Center, is an underground fortress located three miles west of Fort Carson, deep in the granite core of Cheyenne Mountain at Colorado Springs, Colorado. "It is from this facility," said Douglas, "operated jointly with Canada and manned around the clock, that in case of air or missile attack against Canada or the United States, planes and missiles would be dispatched for our continental protection." The firm designed it and supervised its construction.

When first conceived in 1959, the project became the focus of a bitter jurisdictional dispute between Army and Air Force engineers. In haste, the Omaha District of the Army Corps retained the firm to study the advantages of different sites. One suggestion was the Air Force Academy; another, an unidentified spot for a cut-and-cover installation; and the third, seemingly the most secure, the granite core of Cheyenne Mountain.

Exploratory borings were made over the eastern slope of the mountain, and from photographs taken in the bores it appeared that all the joints ran in one direction. By orienting the caverns at right angles to them, the engineers hoped to reduce the hazards of spalling and to avoid the sheared rock zones with which the granite was known to be laced. However, once they got inside the mountain and the large caverns had been substantially blasted out of the rock, they discovered that "one shear zone curved like a scorpion's tail, and struck diagonally across the intersection of two main chambers"—right where the nerve center of the command post was to be. At that juncture the rock was highly weathered and weak, and there was considerable doubt as to its integrity even for long-term static loads—to say nothing of its resistance to the effect of a blast. This threw the construction into crisis.

The only obvious solution was paradoxically out of the question—a reinforced concrete lining, supported by rock bolting, that would have made the roof and walls prohibitively thick. On the other hand, long rock bolts alone would not suffice. There was thought of abandoning the project altogether, or reconstructing it anew in another configuration. "Out of that crisis," according to Kuesel, "the great sphere of Cheyenne Mountain was born." This was a 100-foot diameter concrete sphere, "with cylindrical appendages extending into the four adjacent chambers—a grapefruit with four tin cans attached. The intersecting spherical and cylindrical surfaces mutually reinforced and supported each other, and transferred the load from weak to solid rock."

The solution, however, presented new construction problems. It was necessary to enlarge the span of the excavation by half—increasing anxiety among the miners who were already uneasy working under the fragile roof. And so one afternoon, in a dingy cafe across the street from the Corps of Engineers headquarters in Omaha, Douglas and Kuesel came up with the design of a central steel tower support that would make reinforcement of the weak rock zone practical. Subsequently, six heavy steel columns laced together were jacked against the center of the rock dome to provide temporary support while the sphere excavation was completed. However, only Douglas and Kuesel knew that the tower could only carry about 5 percent of the weight of the rock, and that its chief benefit was psychological.

§

The NORAD complex is made up of six chambers. The three largest, parallel to one another but divided by pillars 100 feet thick, are 60 feet high, 200 yards long, and house three-story buildings. The others, only slightly smaller but a little farther apart, are connected by tunnels to the first. There is an L-shaped access tunnel from the surface northeast of the complex, and air and exhaust tunnels leading out of the mountain to the south. Support facilities include a diesel generator and a water reservoir, while sewage disposal is by an outfall line to a treatment plant to the east—all so those within can "live for a considerable period of time without going outside."

Some features of NORAD are, to say the least, eccentric. For example, the buildings are "shock mounted" and sit on helical wire springs four feet high, the "metal of which," said Douglas, "is bigger around than my forearm." These are to allow the buildings "to vibrate independently for a short time after an attack."

To shield electronic instruments against the electromagnetic pulse effects of a nuclear blast, the buildings have an outside lining of thin low-carbon steel plates. In certain places, such as stairwells, which lack a cut-off wall, a kind of airlock is installed with two sets of interlocking doors, neither of which can be opened unless the other set is closed.

So shielded, the buildings acquired a monocoque design—like the hull of a ship, only with the chamber at the top instead of the bottom. "Upside down ships," one engineer called them, "constructed in a cave and floated on a sea of springs."

The steel plates were the idea of Nate Newmark, an Illinois professor known to his colleagues as "Mr. Underground." He wore a wristwatch with an alarm on it, and when you went in to see him, "after fifteen minutes it went off and your time was up."

During this period, both on and off the project, Douglas proved himself a man of daring. A colleague remembers a terrifying trip from Chicago to Urbana crammed into the back seat of a Piper Cub air taxi with Douglas up front next to the pilot. "We were bouncing like a cockle shell in the thermal updrafts over the Chicago suburbs, and then the pilot turned the controls over to Walter, who proceeded to fly us down to Urbana, though he'd never had any flight training or flown a plane before."

Such bravura was in character. He once said: "I never did anything for which I had the training. There's no use saying 'I didn't take a course in it.' You have to take the attitude that you have all the fundamental knowledge you need to acquire any specific knowledge in the general field." Said a colleague: "Douglas brought fundamental knowledge to bear on new things."

NORAD was completed in 1964, but in the meantime Douglas had undertaken something else for which his specific training was, if anything, less.

Like NORAD, it was *sui generis;* and he regarded them together as the twin peaks of his career.

CHAPTER 21

I n 1952 the firm was invited to make a feasibility study for a rapid transit system covering the nine-county San Francisco Bay area. This study was the first ever undertaken to determine the role of rapid transit in urban transportation in the automobile era, and Douglas was one of the first professionals to correlate mass transportation planning with land use. From such acorn-like beginnings grew the great oak of BART (the Bay Area Rapid Transit System), probably the largest project in the firm's history and without question seminal in the rebirth of interest in mass transit in the United States.

Douglas would call BART "the tenth wonder of the world."

Norma Westra van Brock, who had joined the firm in 1951 and worked with Ziegenfelder on a master plan for Albany, was placed in charge of the planning task force, and in collaboration with Ziegenfelder and Douglas (with help from Peter R. Vandersloot, Elwyn H. King, H. Dean Quinby, and Harry Moses) she conceived and developed the *Regional Rapid Transit* report, an enormous hardcover volume which projected transit requirements for the 7,425 square mile, 9-county Bay area.

The study was finished in 1955, and included in its recommendations a tube beneath San Francisco Bay. On June 4, 1957, the California State Legislature approved creation of a Bay Area Rapid Transit District to plan, build and operate the system. In May of 1959 Parsons Brinckerhoff Quade & Douglas, the Bechtel Corporation, and Tudor Engineering Company created the joint venture, PBTB, which proceeded to develop the overall plans.

Though six counties eventually withdrew from the scheme, the three that remained—Alameda, Contra Costa, and San Francisco—comprised a sizable region of hills and mountains, flatlands and agricultural valleys surrounding 513 square miles of rivers, straits, and bays. In a referendum on November 6, 1962, the voters approved $792 million of public money for the work. Three weeks later the District retained PBTB to do the engineering and manage the construction.

At the project's height, over one thousand design professionals and technicians were mobilized.

The joint venture created two organizational umbrellas for BART—a Board of Control and a Central Office. John Everson was Douglas' alternate on the Board of Control and was the true field general, while Douglas remained in New York. However, as six months of litigation directly followed the referendum, most of his early battles were in court. In the face of considerable duress, with his deputy, Winfield Salter, he held the joint venture together and dealt with the questions and arguments of the local citizenry and no less than 37 municipal governments. He also had the unenviable job of having to parry insistent challenges from the press.

Everson also headed the firm's San Francisco office which was devoted exclusively to the BART project, with Kuesel as assistant manager of engineering and Peter Vandersloot manager of scheduling in the Central Office. Everson and Salter were in charge of the PB contribution to design (except for the trans-bay tunnel, which was supervised by Bickel and Donald Tanner in New York), while Salter and Rubin worked on surface route location and William Armento and Elwyn King on cut-and-cover and bored tunnels respectively. George Murphy looked after the design of the trans-bay tube, Roger Stevenson the construction; William Pease and Harold Wombacker made sure the tube sections, fabricated by Bethlehem Steel, were placed properly in the bottom of the Bay.

The BART system was laid out in the form of an "X," with Oakland at the hub. The legs of the "X" ran southwest through San Francisco to Daly City, northwest to Richmond, northeast to Concord, and southeast to Fremont. In terms of construction, this configuration resolved itself into 75 miles of double-track railroad—a third at grade, a third on viaduct, and a third underground. The underground portion, easily the most challenging, consisted of nine miles of cut-and-cover subway (including 17 stations); 14 miles of single bore tunnel; a three-mile twin rock tunnel through the Berkeley Hills; and a three-and-one-half-mile tube tunnel under San Francisco Bay. The major portion of the underground work fell to the firm to do.

If the pivotal problem posed in the construction of NORAD was solved in a dumpy cafe in Omaha, basic planning for BART was determined in a back room at Bernstein's Fish Grotto in San Francisco, where Everson regularly lunched with his associates from Tudor and Bechtel. A couple of days after the BART bond issue was passed, "we roughly sketched out on paper placemats how to divide things up—so many millions here, so many millions there, so much of this, so many people for that, and who would do what. We used our best judgment of course; but we pulled it out of the air. Years later we looked at it, and though the job had grown immensely in size and complexity, its overall shape was about the way it came to us around the luncheon table that day."

Overall, the idea of BART was "to unify a Balkanized area": in one combined system, to collect suburban passengers from outlying areas in the manner of a commuter railroad, and bring them to their down-town destinations "in the manner of an internal distribution system" while at the same time tying various centers of the region together.*

BART would eventually cost $1.4 billion, but the bond issue did not account for it all. For example, the Trans-Bay Tube was an additional expense; and as the project advanced the dimensions of some of its basic appendages grew. "Everyone wanted something extra," says Everson:

> a bigger station, a facade that was more ornate—and the political and other pressures were too fierce to resist. Despite our protests that money would run out (it was already committed) they couldn't see it because it looked as though there were so many hundreds of millions left. Even items amounting to relatively minor percentages resulted in very large dollar amounts. We let out $50-60 million contracts like they were going out of style. Then came that terrible Christmas when we received the huge bids on the Oakland subways and the Tube.

It was well understood that the deep-level subways planned for downtown Oakland would require difficult cut-and-cover construction. Nevertheless, the low bid was 30% over the Engineers' Estimate,

---

*BART unified the region, but also profoundly changed it. "It was like playing God," remembers a long-time employee. "We'd choose the location of the Berkeley Hills Tunnel and thereby decide which valley was going to develop and which was going to wither. There were major impacts on the pattern and growth of the area. When I went to San Francisco in 1962, in the whole city there was one modern building, the green glass Crown Zellerbach Building on Market and Bush streets. And it was designed and constructed to face a side street. In the succeeding 20 years there's been a tremendous building boom, and it's certainly significant that no major building has been built more than a five minutes' walk from a BART station."

which raised questions in public and official circles as to the engineers' credibility for design and budget control. Moreover, "it was feared that if the construction industry found it could force BART to abandon its budgets, the effect on the remainder of the work would be financially disastrous." Accordingly, it was decided to reject the bids, carve the work into six separate contracts, and redesign the two major stations to simplify their construction.

This had to be done in six months, or about half the time normally required for such a task. Nevertheless, six months to the day from the time the original bids were allowed to die, the redesigned work was advertised. Thirty-one bids on the contract were received, and the sum of the lower bids came within the original budget—and $13 million under the original bid on the single contract.

The crisis in Oakland had meanwhile inspired a sober cut-back in the plans for San Francisco, where certain station designs were simplified. Even so, the construction challenges were such that "after several re-readings of the engineering contract," remembers one engineer, "there was a strong movement to underpin half of San Francisco." In Oakland and Berkeley as well as San Francisco, the tunnels were driven beneath congested streets, through a wide variety of soils ranging from dense, cemented sands to soft, plastic clays. Nevertheless, 20 miles of subway construction were completed with only two major underpinning jobs and without a single significant claim for damage to adjacent structures. Though there were some close calls. In one instance, "a pocket of running sand suddenly flowed into the heading, opening up a chimney to the surface directly beneath an operating streetcar track." Elsewhere,

> a thriving hardware store occupied a building constructed on top of the remnants of several previous buildings destroyed by fire. The structural support for part of the first floor consisted of deeply charred wood beams resting on piles of loose bricks roughly approximating columns. Despite a recommendation to demolish the structure, it was decided to leave it in place, only strapping the loose bricks together, while two tunnels were driven directly beneath the basement floor, one with a soil cover of only seven feet above the crown.

The tunneling under San Francisco's crowded Market Street was 75 feet down in saturated soil. The site "encompassed the worst geological conditions in the city—an area of soft mud that had been filled during

Gold Rush days, and which was littered with the buried remains of sunken ships, debris from the 1906 earthquake, and the remains of old cable-car railways." The water table was lowered by sinking wells; even so, compressed air was required to aid boring by shield. Some stations lie practically submerged, their massive structures contending against the hydrostatic pressure that would otherwise cause them to float. The largest, in fact, compare in volume to ocean liners and in a sense were designed like ships: watertight, carefully weighted, with their interior structure kept light.

But the Trans-Bay Tube was undoubtedly the most flamboyant feature of the system; certainly it aroused the most curiosity. For a start, it was to be the longest, deepest subaqueous tunnel in the world; just as striking, it had to be built across a long seismic fault block, bounded on the east by the Hayward Fault, and by the San Andreas Fault on the west. It was the San Andreas Fault that had caused the catastrophic earthquake of 1906.

To prepare for the worst—that is, for an earthquake comparable to the tremor of 1906—the tube would have to endure a displacement of two feet transverse in a length of 2000 feet; it would have to be elastic to survive.*

Of two-track, binocular design, the Trans-Bay Tube is made up of 57 sections, each 48 feet wide, 24 feet high, and 330 feet long, for a combined length of 3.6 miles. In the customary manner, the sections were temporarily bulkheaded, launched, outfitted with interior concrete linings, towed to the appropriate site, and sunk in place in a trench dredged in the bottom of the bay. Underwater, they were joined with the help of divers by railroad-type couplers which brought a ring of rubber gaskets at the end of one section to bear against the flat end of the next. Water was then drained from the joints, the bulkheads removed, steel plates welded across the joints from inside, and the tubes covered over with sandfill dumped from above. Their

---

*However, some comparatively prosaic problems had to be disposed of first. As it happened, the ideal tunnel alignment coincided with an interlacing network of underwater cables, which therefore had to be relaid; while some very peculiar ideas from people "high up" in the region were obliged a solemn hearing. One of the most remarkable was to take all the mothballed naval vessels anchored at the Mare Island Base and sink them at intervals across the bay. They could then, it was suggested, be welded together, cut through from ship to ship for tracks, while their smoke stacks sticking out of the water would supply the tunnel with air.

What a smokestack-studded bay would have meant for navigation is easy to guess.

alignment was controlled by a laser beam projected across the bay from the shore. At night, the beam was visible as an orange light for seven miles.

The 50th tube was the hardest to get right—far out in the deepest part of the bay. It had to be adjusted; but when it was winched a little it wouldn't come up. When winched further it began to pull the barge down. A diver plunged to investigate, and as he came around to one of the bulkhead doors he spotted the tail of a fish sticking out of the access hatch. The gasket cover, rolled or damaged, had opened a leak, and the fish had wedged itself in. As a result, the entire tube had filled with water. To pump the water out (2.5 million gallons in all) another cover was welded over the hatch plate and fitted with valves.

It is said that in the event of an earthquake the Trans-Bay Tube will be the safest place to be. This is possibly true. Some sections are curved to conform to the bottom of the bay, and all, in a sense, float in alluvial soil in relative isolation from the shocks that would be transmitted through solid rock. At either end, sliding joints, which allow movement in any direction, join the tube to the more solidly anchored structures on shore.

CHAPTER 22

In many respects the engineering for BART was brilliant. But it was widely believed that it had to be something more. To lure people out of their cars, it had to appear downright exotic; to capture their imaginations, it had to be "beyond the state-of-the-art." From this requirement came its space-age look, its streamlined cars and fully computerized and automated character. Like rockets or missiles, the cars are remote-controlled from a "situation room" furnished with consoles that monitor display boards flickering with colored lights. Central computers control pace and scheduling; wayside electronic gear control stopping, starting, and speed.

Considerable attention was also given by a small battalion of architects to avant-garde and colorful design in the stations, which were optimistically conceived as cultural rallying points for the local communities. Clearly it was hoped that not only in a technological but aesthetic way the system might serve as the model for rapid transit in the new age.

In this it did not exactly fail; but it did not exactly succeed either.

Technologically, at least, "once launched BART behaved like a neurotic machine. Trains stopped inexplicably between stations, speeded up instead of slowing down, slowed down instead of speeding up, glided past stations where passengers were waiting to board, and stopped at stations without opening their doors."

On one occasion a trainload of passengers was shunted off the line into a yard. "Traffic was picking up," remembers Everson, "and an additional train was put on the tracks. However, nobody bothered to inform the computer—which, in computer language, kept asking the train, 'Who are you? Who are you? Where are you going?' and all it got back was static. Finally, it decided to get this unidentified thing off the line, and at a convenient place—a train yard—it did."

Moreover, "ghost trains" soon began to show up on the control boards, while others actually on the tracks disappeared. "The blame for all the difficulties," says Douglas, "is pretty widely placed."

I sat on the Board of Control for the joint venture, and at a point in the work the District began building up its own organization. They would move in to take over shop inspection and fabrication of the equipment, and a great deal of the train control. And I thought at the time we should not allow that, that should be protested. But I didn't speak up. The joint venture ought to have written to the District saying we would like our position clarified so that we have full responsibility for those matters for which we are responsible, and are not responsible for what the District wants to take over.

However, by 1972 public and political pressure forced service to begin prematurely, and the six-month shakedown period, provided for in the original schedule, was dropped. The ribbon was cut, and all the debugging was done in service, and on page one of the newspapers. Thus, when the first line in East Bay opened, rust on the rails occasionally prevented completion of the electronic circuit by which the individual trains were tracked. "Many of the same people," says Everson, "who had insisted on something different and wonderful began asking, 'Why did you build this stupid thing? You could have bought something reliable off the shelf.' " And the District paid heed. It filed a suit against both the engineers and the suppliers, who in turn, filed counterclaims. Eventually the matter was settled out of court, though while it lasted, according to one principal, "it lurked around us and had the potential for ripping us apart."

If the system lost some of its glamour and gloss, most of the bugs are out of it now, and it appears to be working well.

CHAPTER 23

T hroughout Douglas' partnership career, his principal colleague
was Alfred Hedefine. They had the same partnership interest,
the same shareholdings in the company, and might therefore
have been expected to hold equal company rank, with the difference
that it had always been understood that Douglas would be the business and administrative head of the firm. In fact, when the name had
been changed to "Parsons Brinckerhoff Quade & Douglas," the addition of Hedefine's name had been discussed. But "that got to be a bit
much" so it was implicitly agreed to adjust it—to Parsons Brinckerhoff
Douglas & Hedefine—when Quade retired. Quade died on the eve of
his retirement; but the name was never changed. Nor was the subject
ever raised. "Hedefine," says Kuesel, "was simply too much of a
gentleman to bring it up."

Alfred Hedefine was unquestionably a great engineer, as a technical professional aptly to be compared with Quade. Born in 1906, the
son of a power plant engineer, and educated at Rutgers University, he
took his first job in the midst of the Depression operating a pile driver.
In 1933, Waddell & Hardesty hired him to work on the Mill Basin
bascule bridge of the Belt Parkway in Brooklyn, where he developed
innovations that became the basis of his master's thesis for the degree
of civil engineer from the University of Illinois. The Mill Basin Bridge
was followed by the design of the Marine Parkway vertical lift bridge,
also in Brooklyn; the Rainbow Arch over Niagara Gorge, then the
longest hingeless arch span in the world; and the St. George's tied
arch bridge over the Chesapeake and Delaware Canal. In 1938, at the
age of 32, he designed the famed Trylon and Perisphere theme buildings for the 1939 World's Fair.

During the war he served with the Eighth Air Force in England, in
intelligence and photo reconnaissance work, and after returning to
Waddell & Hardesty (now called Hardesty & Hanover) moved on to
Parsons Brinckerhoff in 1948 as head of the Bridge Department.

A very distinguished-looking man, and something more than an
amateur organist, Hedefine was cultivated, courtly, reserved, and

dapper in his dress. Decorous toward others, he had a strong sense of his own professional worth and expected in turn to be treated with respect. "His demeanor gave you some fear about him," recalls a colleague, "that you might not handle yourself correctly, in his judgment."

In a sense, just as Bickel succeeded Thoresen as a master of tunnels, so in bridges Hedefine succeeded Quade. But to put it that way would not be quite right. Hedefine and Quade had been peers all along, and Hedefine "succeeded" him simply because Quade retired.

The Arthur Kill, Myrtle Avenue, and Talmadge Memorial Bridges were chiefly his work, as well as the Martin Luther King, Jr. Memorial Bridge in Richmond and the 62nd Street Bridge in Pittsburgh, both of which won awards. More celebrated was his double deck Fremont Bridge in Portland, Oregon, a slim, graceful structure conceived as a "carefully balanced spring" which set a continuous beam against the compressive thrust of the arch "in the manner of a bowstring tying the ends of a bow." A bridge of three spans, its central span of 1255 feet made it the longest tied arch bridge in the world.

Moreover, Hedefine knew quite a bit about tunnels, and with Bickel prepared studies for a railway tunnel under the English Channel, and for an immersed tube tunnel crossing, supported on an underwater rock dike, for the Straits of Messina. The latter, submitted to an international design competition, won an award.

In other areas, he prepared uniform criteria for design and construction of all U.S. military airfields, developed a unique concept for an automated vertical storage-and-retrieval system for handling large shipping containers, and supervised a comprehensive urban transportation study for Hamilton, Ontario, the second largest city in the province and a commercial crossroads of the Great Lakes.

However, Hedefine's outstanding project by far, and the culmination of his diverse career, was the Newport Suspension Bridge over the east passage of Rhode Island's Narrangansett Bay. If he had done nothing else, his principal role in its creation alone would have been enough to secure his name for posterity. By consensus, it is one of the most beautiful suspension bridges of modern times.

For years the bay had effectively severed any direct and continuous route between southern Massachusetts and points along the coasts of Rhode Island, Connecticut, and New York. Moreover, the ferry service stopped in violent weather (which was frequent) and seldom continued through even the most halcyon nights. The Jamestown Bridge built in 1940 over the west passage had helped; but a companion crossing to the east was long in coming, despite annually mounting witness to its need. The through traffic from New York and Connecticut kept growing, while jazz, folk, and other festivals drew people in flocks to Newport and other havens about the bay.

Between 1940 and 1965, thirty-five locations and span configurations were considered—most of them sound enough, but all rejected out of hand by the Navy. Hedefine was involved "in all the twists and turns of this road."

The Navy objected to a fixed structure because the passage was constantly used by all their vessels, particularly the submarine fleet, administered from the Newport Naval Base. Accordingly, every two years or so Hedefine would visit the new commandant, brief him on communications problems around the bay, including the difficulty of getting men and materiel from Quonset to Newport by way of Providence, the traffic delays, medical problems in emergencies, and so forth. The admiral would listen, call for the file, read his predecessors' objections, add a few more of his own, and return the file to the drawer.

Then one night a ferry carrying Mamie Eisenhower across the bay through a dense fog was nearly sliced in two by a destroyer. "The next morning, the admiral received a telephone call from 1600 Pennsylvania Avenue—and the Navy's attitude changed remarkably. They found they could live with a bridge after all. It really ought to be called the Mamie Eisenhower Bridge. Without her it might never have been built."

Hedefine's bridge design appeased the Navy's fears. It allowed for a wide horizontal clearance and a height of 215 feet, enough to permit passage of the tallest ships afloat. In this respect, it is exceeded in the United States only by the Golden Gate Bridge. Its other distinctions include the pioneering use of prefabricated, parallel-wire strands, a significant advance in bridge-cable construction; new procedures for driving underwater piles at depths of up to 162 feet, the deepest ever attempted, and for placing the largest amount of concrete (90,000 cubic yards) ever underwater; finally, it marked the first use of "rocket-launcher" anchorages, which transmit the cable leads to the rear of the anchorage in a compression rather than a tension mounting.

The use of prefabricated, parallel wire in the cables meant that the stranding could be done intermittently over a period of several months at a plant while the towers were being erected at the site—as opposed to the conventional helical strand method, or Roebling's cable-spinning method of looping two or four wires at a time over a wheel that carried them from anchorage to anchorage over the towers. The latter was a hazardous procedure, and required standing on a wind-buffeted catwalk hundreds of feet in the air. Indeed, the history of hurricanes and swift floodings in the bay prompted Hedefine to take special pains to protect the footings of the bridge from scouring and its superstructure from gales.

The soundness of the design "received an unexpected tribute when in February 1981 a 50,000-ton oil tanker collided with one of the main piers, scoring a direct hit in a dense fog. The bow of the ship was shortened ten feet by the impact, but the bridge did not budge and the only damage it suffered was an enormous blotch of grey paint spread over the end of the pier."

A bridge of three suspended spans, its proportions are exalting to the eye. To balance the long approach spans from the shore, its 1600-foot main span has two side spans each 688 feet long; while the Gothic arch-like top strut of the towers is pointedly reminiscent of the Brooklyn Bridge. Reminiscent, but a historical reply: Ahmet Gursoy, who was a principal designer of the superstructure, observed that the bridge was "meant to symbolize easy passage between two points" —for Hedefine the quintessence of modern bridge design. In 1960, Hedefine wrote:

> The slower rate of past ages was more productive of things of permanence, and so this quality of "permanence" was found to have been developed to the intensity of an ideal. Of necessity this ideal regulated the spirit and the manner of the growth of bridge building. Now, today, this ideal has given way, without suffering, to another ideal which can perhaps be best embodied in the term "speed." Permanence is a conquest of time; speed is conquest of space, and today bridge building is obeying this universal impulse to speed. The bridge of today represents lightness and economy, against the mass and abundance of the past.

Others who contributed significantly to the bridge were Herbert Mandel, as project manager, and Lou Silano, as project engineer.

The Newport Bridge opened in 1969. But its magnificence was marred. Simply put, the paint was peeling off.

This was not Hedefine's fault. Together with John White, head of the Specifications Department, he had developed a new epoxy paint system that would, and should, have protected the bridge for many years both from the salt of the bay and from de-icing salts. Unfortunately, the surface of the steel was not properly cleaned, and a great many occlusions of debris were trapped in the paint. Hedefine saw from the beginning what the problem was and refused to approve the plans of the contractor to put the bridge up. But he put it up anyway, without correcting the defective paint job, and it affected the whole superstructure, especially the suspension spans. The contractor said, "We'll argue about the paint later." And they argued about it for a great many years.

Predictably, the contractor claimed that the paint was experimental and inadequate. Hedefine rejoined that there was nothing wrong with it, the trouble lay in how it had been prepared and applied. Eventually, another contractor with a sandblasting crew removed the old coating and with the same epoxy system painted the bridge anew.

The paint held up; moreover, after the longest trial in the history of the State of Rhode Island, Hedefine was vindicated in court.

The controversy, however, had taken a great toll on Hedefine's health which, under the best of circumstances, was precarious. He was a diabetic, and from a misdiagnosed case of appendicitis in his youth, peritonitis had eaten away most of his abdominal wall. He wore a supporting corset, but his intense professional drive further undermined his strength. A long commute from his home at Lake Mohawk, New Jersey together with his habit of working late, often obliged him to stay overnight in hotels. "He never took a vacation," says a colleague. "He didn't take care of himself, he worked himself into the ground." Soon after he retired in 1971, he developed severe circulatory trouble in his legs, and on January 26, 1981, he died.

Hedefine's loyalty to Rutgers, his alma mater, was akin to Parsons' devotion to Columbia, and at his passing the university flew its flag at half-mast. In a moving eulogy, Kuesel declared: "We will not see his like again."

CHAPTER 24

Douglas' tenure was not untroubled. BART possibly diverted a disproportionate percentage of the firm's talent to one endeavor, and encouraged "sloppy business practice" in general precisely because it was so lucrative. Moreover, though years before Douglas had properly criticized Covell for delegating so much, in his own determination to remain completely in control, he appears on occasion not to have delegated enough. "He didn't," recalls one principal, "always pay enough attention to his own jobs." Finally, his high-handed if unintentionally abrasive manner left some lasting scars. Nevertheless, the Douglas era set the firm on a new course. With BART it was again at the cutting edge of rapid transit technology. Just as William Barclay Parsons' cut-and-cover IRT had set precedents that lasted for fifty years, BART set the new precedents.

In 1961, Douglas also spearheaded the study for Atlanta's MARTA, the first rapid transit system in the southern United States, and directed transit studies for the regions of Philadelphia-Camden, Pittsburgh, Baltimore, Chicago, Detroit, and St. Louis. Abroad, he outlined a rapid transit system for Caracas, Venezuela.

Meanwhile, the firm continued to diversify. In water supply, programs were developed for the Greater Cleveland area (using new computer techniques) and for Denver by way of 62-miles of collector tunnels discharged into Dillon Reservoir and fed into the 23-mile Roberts Tunnel through the Continental Divide. Airport work included expansion of the Jamestown Municipal and Atlanta Airports, the conversion of Olmsted Air Force Base in Harrisburg, Pennsylvania to a commercial airport, and a study for making the Southwest International Airport into a regional hub for the Dallas-Fort Worth area. Abroad, there were studies for the expansion of air bases in Okinawa and Thailand and in national economic planning for Colombia, Equador, and Peru. In highway design, several extensions of the New York State Thruway should be noted, along with construction of new highway connections from the Garden State Parkway to the seashore resorts of Seaside and Avalon in New Jersey; in waterfront work, a mile of

continuous bulkhead berths for the Port of Toledo, an integral plan
for the ports and waterways of Peru, and wholesale improvements
which doubled the size of the Port of Callao near Lima.

There was also the 63rd Street East River Tunnel for subway and
Long Island Rail Road riders between Manhattan and Queens. This
last project (though not yet in service) was landmark. It included the
first four-track, double-deck sunken tube tunnel ever constructed,
and seamlessly joined three distinct kinds of structures: two bored
rock tunnels to the tubes, and the tubes to a midriver Roosevelt Island
Station mined out of rock. The transitions were the crux: instead of
normal cut-and-cover connections in the dry (using cofferdams) the
tube sections had to be connected directly to the rock tunnels
underwater, and "this had never been done before." Here it had
to be done four times. The method was "to place each tube section
with one end in a niche close to where the rock tunnels would be
holed through; fill in the space between with tremie concrete; and
then tunnel through the remaining rock and the concrete 'plug' to
the tube."

§

Inevitably, there were projects which failed to come to pass, among
them a tunnel under the Tagus River in Lisbon, a hydroelectric plant
for Ankara, Turkey, and a two-lane vehicular tunnel between Finland
and Sweden. Another was a long rock tunnel from Caracas to the
Caribbean Beach, ostensibly for general vehicular use, but perhaps
intended as an escape route for the dictator Jimenez and his lieutenants in the event of a coup. "I got my first intimation of the atmosphere in the country," said Bickel, the partner in charge, "when I
landed in Caracas and the customs inspector very carefully scrutinized my copy of *Time* magazine." His colleague, George Murphy,
was similarly importuned on another trip. Arriving in Tokyo for a
tunnel conference with a film on the Trans-Bay Tube, and carrying a
case of bourbon as a gift, the bourbon aroused suspicion and the film
was briefly confiscated as pornography. (The inspectors unwound it
by hand.)

§

One project that went awry might have cost the firm a great deal.
But Douglas put his best foot forward and turned it around. As it

happens, the incident also illustrates how, as he said, he "lived the business 24 hours a day." One Thanksgiving, as he was about to carve the family turkey, the telephone rang and it was his resident engineer on the Port of Toledo.

"Mr. Douglas," the man said, "the bulkhead's fallen in." I handed the carving tools to my brother-in-law and said, "I'm leaving." The firm had in fact made a serious mistake in the design, and I told the client. The contractor had also left off some welds. I called our insurance company and explained the situation and said, "I believe if I have your backing I can negotiate to pay for material that should have been in there anyhow, and I'll negotiate with the contractor to see how we can divide the responsibility if I have your authority to do this." They said, "Go ahead." And I negotiated with the contractor and he agreed to pay for the labor and equipment and I agreed to supply the material that had to be replaced. So the costs were divided three ways. The insurance company was so pleased with the outcome that they reimbursed us for the total cost of our redesign, and our client became a stronger client than ever because I had never lied to them or tried to obscure the facts. So we did all the rest of their work. And we had a mile of bulkhead on that port.

This was not untypical. It could be said of Douglas that he was a man who believed that honesty was the best policy and that it usually paid off.

So the firm grew. Under Douglas it expanded its architectural abilities, bought a computer, and in addition to the design and management of public works of all kinds, developed two new disciplines: transportation planning on a regional and national scale; and the analysis of construction with respect to its impact on the environment, or "environmental technology." The last was overdue. As Arthur Jenny, one of the firm's leading highway engineers remarked: "In the old days we didn't worry about environmental considerations. We just put the highway through."

In 1966 the firm also established semi-independent affiliations with two outside companies: the National Electric Service Corporation, utilities consultants; and Lord & Den Hartog, Architects and Engineers. Both were soon absorbed into Parsons Brinckerhoff. Perry Lord became a partner, assuming direction of the firm's outpost in Boston where with Greenfield's help he obtained considerable work through the Massachusetts Bay Transportation Authority and the Department of Public Works. Lord's expertise in architecture also made the Boston office a center for the firm's architectural activities.

Looking to the future, Douglas predicted that rapid transit would supplant highways as the lion's share of the firm's profitable work, and prudently advised a modest (25 percent) increase in foreign work but "no more: we become vulnerable to political changes." Among immediate domestic prospects, he singled out the New York West Side Highway project, or Westway, as a particular plum: "If the project materializes in actual design, we will be immensely fortunate because that work will extend over a decade." Overall, it seemed to him the important thing was to do "much more work for private enterprise clients," since they were prepared to pay "at the very least 1.85 times payroll costs," as opposed to the less inviting cost plus fixed fee compensation for public works.

Thus spoke Douglas the businessman. On the eve of his retirement, in 1975, the firm was billing over $15 million a year and had branch offices in nine cities, both here and abroad. He concluded proudly: "We are diversified to a point where we can staff large-scale programs almost entirely with our own personnel."

As a professional, Douglas' contribution was impressively acknowledged by the Construction Division of the American Society of Civil Engineers, which named him "one of the top ten construction men in the past half-century"; and in 1970 he received the non-member award of the Moles, the accolade (as given by his peers) he values most.

CHAPTER 25

E xeunt Macdonald's appointed guard. Enter the bright young men of the late 1940s and early 1950s, who by 1975 had come of age: Greenfield, Dyckman, Kuesel, Salter, Rubin, and in 1965, Henry L. Michel, an engineer with an international background who would profoundly alter the character of the firm.

From the very beginning, the organization of the partnership had resembled the Articles of Confederation. There was a first among equals, of course, and an order of seniority, but otherwise each of the partners enjoyed eminent domain with respect to his clients, projects, and staff. They shared office space together, but they signed their own contracts, sent out their own bills, and collected their own receipts. The receipts were pooled, and after costs were retired, ownership interest in the company determined what each took home.

In the Douglas years, however, the firm began its circuitous and rather peculiar evolution into a corporation.

Loosely speaking, Macdonald had prepared the way: in the early 1950s, he created for the partnership a corporate identity to meet the licensing requirements of certain states. But this was a nonprecedent, for the corporation (Parsons Brinckerhoff Quade & Douglas, Inc.) did not at first practice engineering. It was merely a device for business administration. The employees (and partners, in addition to their partnership interest) were on a corporate payroll, which made it easier for the firm to handle fringe benefits such as medical, bond savings, insurance, vacation, and sick leave. The corporation owned the furniture and equipment and defined the overhead. But its principal business value seems to have been as a repository for various funds that as they built up increased the book value of the stock.

Meanwhile, in response to an oddity of California law, another corporation had been expressly created for BART. The preliminary planning study had been undertaken by the partnership, but when the contract graduated to heavy design, the partnership decided it had better have the limited liability of a corporate shield. However, the partners found they could not assign the contract to the New York corporation because California law decreed that in order to register the corporation, the principals named in the title had to have licenses to practice engineering in the state. Quade and Douglas had them, but Parsons and Brinckerhoff did not. Nor could they get them: they were dead.

Douglas rummaged through the company files, came up with a New Jersey corporation, and changed its name to PBQ&D, Inc. That satisfied California law because it was a nameless entity. As Douglas remarked, "if we had called ourselves 'Rascals, Inc.' that would have been all right, too."

Now, PBQ&D, Inc. was at first a single-purpose corporation, and because it had only a few shareholders it turned out the corporation could be registered under Federal Subchapter S as a Small Business Corporation. This meant that it could distribute its earnings annually as a partnership would. Thus, all the revenues of the BART project, which were considerable, flowed directly to the partners. However, its liability shield was imperfect: the BART district insisted that the partnership stand behind PBQ&D, Inc.'s performance, so when litigation indeed came about the threat to the partners' personal assets was great.

For a considerable period during the Douglas era the business still operated as a partnership. The accounting department ran a corporate accounting setup, but the decisions were made at meetings of the partners. The decisions were made, that is, by Walter Douglas, with the other partners nodding in agreement. Not quite so autocratic as the Macdonald era, but Douglas was very definitely the dominating force.

So the firm was a corporation, but not a corporate business organization; and the partners held on tenaciously to partnership ways.

Everything changed in the latter days of Douglas' reign.

There was a crisis in the succession, unprecedented in the history of the firm.

What happened?

In the first place, Douglas had been ill, and his capacity to lead sufficiently impaired to require a sort of informal government by committee—though in matters of substance his say was still law. Second, a number of other partners were fairly close to retirement age, while the remainder were comparatively young, a fact which may have encouraged in Douglas a certain "serene conviction in the incompetence of his successors." "So," says Kuesel, "he sort of left the future to sort itself out."

Sort of; but not quite. Despite his optimistic predictions about the future, he was apparently not entirely convinced the company's prosperity would survive his retirement.

One possibility he leaned towards was acquisition by a larger firm. Parsons Brinckerhoff was duly approached—by the Dravo Corporation—discussions became earnest, and among the questions Dravo asked was: "How are you organized?" The answer (so far as Dravo was concerned) was: "Not very well." "We looked at ourselves," says Michel, "and decided that because of the risks involved in the larger projects; the lawsuit in San Francisco; the fact that one judge's ruling could wipe out the partnership and the partners personally; and the fact that there was a clear need for a unified leadership, instead of what amounted to nine different companies, each one headed by a different partner and working in a vacuum without utilizing the total resources of the firm—all these things led to a lot of soul-searching. In order to get the most money from the sale we were forced to reorganize, to present them with a package that was both attractive and neat."

In 1974 Michel volunteered to develop a new organizational structure, which featured an executive hierarchy, or "centralized corporate headquarters group" including a board of directors with "command and line functions decentralized" to nine regional activity centers, with 19 offices, including 6 overseas. The overseas offices would be separately managed by Parsons Brinckerhoff International, Inc., a subsidiary corporation.

The executive hierarchy was the critical feature, and alternative schemes were put to a vote. One called for a president/chief executive officer, an executive vice-president in charge of business development, and an executive vice-president in charge of finance and administration—a tripartite scheme. The second inserted a fourth position among these three: a chief operating officer under the president. Although there is some question as to which plan was actually favored, the tripartite scheme prevailed.

All partners not elected to the hierarchy were to be redesignated senior vice-presidents.

Dravo or not (the partners voted against the sale), the decision to reorganize was probably inevitable. "The firm, along with the entire professional services industry," says Michel, "was late in recognizing the necessity of applying modern management techniques to its needs. The companies that failed to do so were acquired—something the firm had just barely avoided—became moribund and lost their position in the industry, or disappeared. In our case, we had lost a position of leadership."

So the corporation became a business management entity, and began to operate in a corporate mode. Key directorial and management positions would now be staffed according to function rather than seniority or ownership, with modern management practices taking hold. The perceived need for a layering of controls and a method to provide checks and balances led to the creation of groups staffed by experts in corporate law, finance and insurance. "We used to do everything ourselves," says Michel. "Now we have 'non-engineers' looking over our shoulders." The partners became the principal stockholders in the new corporation but, for the first time in the firm's history, key employees were also offered shares.

Yet the problem of the succession remained.

As architect of the corporate reorganization, Michel was an obvious candidate for the helm. However, others were first to emerge. The logical choice for chief executive by seniority was John Everson, but Douglas, who was chairman of the board, promptly nominated him to be vice-chairman, thus plucking him from the field. The two leading candidates thereafter were Seymour S. Greenfield and William T. Dyckman, each with a substantial following but neither with sufficient votes to prevail. In a compromise, the mantle fell to Michel, with corporate and business development going to Greenfield, and finance and administration to Dyckman. "Most of the partners were pretty nervous about me," says Michel. "I was known as a somewhat free spirit and independent thinker and there was some concern—if not fear—that I was going to run away with this and really screw it up."

Michel was also made president of Parsons Brinckerhoff International, Inc. Thus the chief executive oversight and the day-to-day operation of the firm fell to one man. With a sort of weary pride Michel now recalls, "When the dust settled, I was in charge of everything."

CHAPTER 26

H enry L. Michel is a chain-smoking, silver-haired, cosmopolitan man, polished, suave, and agreeable, with an optimistic nature that appears to be innate. "One of his greatest assets," remarked Dr. Mario Salvadori, his professor and mentor at Columbia University, "is his smiling face."

Born in 1924, Michel became interested in engineering at the age of 14, when not far from his home in Queens, New York he watched the pier and abutment construction for the Whitestone Bridge. During the Second World War, he served in the Army Signal Corps in New Guinea and the Philippine Islands and afterwards attended Columbia University under the G.I. Bill of Rights. As a student, he came under the influence of Salvadori (now James Renwick Professor Emeritus of Civil Engineering), who sought to communicate something of his broad love of learning to the younger man.

After graduating from Columbia, Michel worked on various jobs while studying in the evenings for his professional license exam. After a stint as resident engineer on an oil refinery in Toledo, Ohio, he was sent by M. W. Kellogg to work on a refinery in Edmonton, Alberta, Canada—"a real boom town," he recalls, where with his wife he arranged concerts in a large cattle-auction shed flooded for ice-hockey. "We put plywood down on the ice and erected a little step platform with a piano on top. The Vienna Choir Boys came, and so did Artur Rubinstein, who played in the bitter subzero cold with the fingers cut out of his gloves."

In 1954 Michel went to England for six years to assist in the upgrading of all RAF fighter-bomber stations to NATO standards, which included "igloos" for nuclear warheads, and then in 1960 was invited by Salvadori to manage the Rome office of Panero, Weidlinger, Salvadori, an engineering partnership. The office had recently been enlisted by Walter Gropius to do the engineering for his University of Baghdad but, as Michel immediately discovered, had somehow managed to squander the fee before completing much of the work. Accordingly, he shut the office down as insolvent and decided to open an office of his own.

His first order of business was to finish the University of Baghdad. Though in those days, he says, there wasn't much work in the Middle East—"oil was selling at $2 a barrel; all the money was going to the oil companies and very little sticking to the fingers of the sheiks"—he nevertheless found work in Libya, Saudi Arabia, Nigeria, Kenya, Tanganyika, and Iran. He befriended the Aga Khan, for whom he designed a mosque in Salaam, buildings in Brazil, and the Costa Esmeralda tourist resort in Sardinia. Locally, in Italy, he built factories and plants for RCA International. "So I became a generalist. I've covered literally every aspect of the engineering business."

In 1965, deciding "not to become a permanent expatriate," Michel sold his company to his two partners and returned to the States. "But I was unemployable. Who would hire the former president of an international engineering firm?"

In those days, in fact, very few American engineers had much international background. However, Walter Douglas, whose vision of the future included a man like Michel, hired him, and within four years made him a partner.

Michel rose to the occasion. Put in charge of the defunct water resources department ("an assignment no one else wanted—which is what the new partner usually got"), he quarried the staff for someone expert in the field. He found Paul Gilbert, who ten months earlier had come to the firm from the California Department of Water Resources but had somehow become submerged in the highway department where he was "all but forgotten." Under Michel and Gilbert, the department staff grew from 6 to 35 in a year. Before long they had secured two large contracts against formidable competition. One, known as the Eagle Piney job, involved design of 62 miles of water collection and delivery tunnels through the Gore Range in the Rocky Mountains for the supply of Denver, Colorado and its environs. The other (CORSIM) pioneered in the use of a computer model which simulated the natural stream characteristics and water rights of the Colorado River, and projected the effect of weather, irrigation, mining and other factors on its future behavior and flow. The model represented a scale of river mechanics in simulation never achieved before.

So Michel established himself as an enterprising partner. Yet (he

o the partnership. He was uncom-
als, its airs. He resented the format
gs: "the same questions, the same
it immutably the same—in order of
ter Douglas, Mr. Hedefine, and so
ne day, Michel came into the meeting
, he recalls, "I walked around the table
e partners. By pure reflex, I think, they
it the same time and stuck them in their
didn't know what they were doing. And
all or — at one another and burst into laughter. I
think that single — nged the personality of the company. Since
then the whole atmosphere has lightened. And I think that's the
greatest contribution I've made to the firm."

When Seymour S. Greenfield took charge of business development
and corporate planning it marked the first time such matters had been
identified as a specialized function and assigned to a professional
marketing staff.

Tall and rather deliberate, with a genial if rugged countenance and
a deep bass voice, Greenfield, says a colleague, is a "supersalesman
as well as a first-rate mechanical engineer." Drawn to the mechanical
side of engineering because "machines and rotating equipment seemed
to me more animate and exciting than static things," he graduated
from the Polytechnical Institute of New York and during the war
served in the Navy as a line officer on engines and turbines. In 1947,
he came to the firm as part of the task force assembled to work on the
NATO bases in Newfoundland and Iceland, and still remembers with
fascination Iceland's twenty-four hour daylight and fine lateral mist or
"horizontal rain." Thereafter, he worked on the blast valves and life
support systems for Raven Rock. John Bickel takes credit for hiring
Greenfield; Malcolm Walsh, the chief of mechanical engineering for
the NATO bases, was his first boss and certainly one of his mentors. "I
learned a great deal from Walsh," says Greenfield. "He was a rugged
individualist if ever you saw one, a remarkable guy."

From the firm's beginning, the mechanical engineer had been
slighted. Though Sanford Apt had done yeoman's work on the Sherman
Island Powerhouse and Dam and later on the World's Fair, and Bickel

had contributed enormously to movable bridges and drydocks before his rebirth as a tunnel engineer, many of the mechanical/electrical assignments had been subcontracted out. Moreover, when Bickel was promoted to partner it was as a designer of tunnels; that is, as a civil engineer. Under Greenfield the firm at last began to acquire a broad in-house competence in the field. "We built up the other mechanical skills from scratch," he says; and it was the military work specifically, "where they were as essential as anything else," that thrust them to the fore. Soon Quade and other structural engineers began to sit up and take notice. "Quade and I used to go out to Omaha to the Strategic Air Command bases and he found a whole new world. He found that there was something else besides bridges." Most of the underground installations were lucrative, says Greenfield, who adds: "I had no moral qualms about such work—though my wife did: she said, 'What are you doing?'—because we were never involved on the delivery side. We were always looking to protect. I found nothing distasteful about that."

When Greenfield became a partner in 1964, he was the first mechanical engineer ever made a principal in the company.

In 1966 he established the Boston office and within six months had brought in about $2 million worth of Massachusetts subway and highway work.

Though part of the new leadership, Greenfield flavored business development with an old-fashioned personal style. "Most of the business I've brought in over the years, particularly in areas outside my professional strength, came about because of the confidence the client had in me personally. The client would say, 'I know you, you're going to get the job done for me, that's why you got the job.' I don't know enough about highways and bridges to design them. I couldn't talk to a bridge engineer about technical details, but I could talk to the commissioner and say, 'Look, give me the job and I'll get it done.' And he would say, 'I believe you, you've got it.' That's what we need to do more of."

The third member of the ruling triumvirate, William T. Dyckman, has been called by Martin Rubin "the unsung hero of the transition from partnership to corporation, the glue that held things together."

The tribute seems plausible. Very methodical, precise, and genteel, with a foursquare build and a calm, broad face, Dyckman exudes a certain proverbial Dutch competence one might look to in times of trial. Says Salter, "Dyckman's solid, conservative approach to matters made him a comfortable person to put in charge of finance."

The son of a long-time chief of maintenance engineer for the borough of Manhattan, he studied sanitary engineering at New York University, served in Bickel's squad in Drydock Engineers, and joined the firm in 1945, where his first assignment was checking rigid steel frame bridges for George Vaccaro. In 1947, he worked with Bickel, Douglas, and Greenfield on the Keflavik Air Base in Iceland, after which he was invited to form a highway bridge design group that eventually grew to 65 people and produced plans for hundreds of bridges and highway structures for the toll roads and expressways the firm designed between 1949 and 1959. In 1959, when Everson, who had been managing highway projects for Bruce and Quade throughout New Jersey and Virginia, was summoned to San Francisco for BART, Dyckman replaced him. For some time thereafter, before BART got under way, Bruce and Dyckman together handled "well over half the firm's work, so far as productive cash flow was concerned." To that was added the management of highway and port work throughout Central and South America, and when Ziegenfelder retired, Dyckman inherited his New York State projects as well. By then a partner, he went on to become principal-in-charge for Westway, for which the firm was awarded the management of engineering in 1972.

To these three—Michel, Greenfield, Dyckman—must be added an unofficial fourth: Thomas R. Kuesel, who regards himself (and is so regarded by his colleagues) as "the professional conscience of the firm." "I went to my partners at the time of Douglas' impending retirement," he remembers, "and told them that in my view the firm would have to have someone to look after the engineering performance and that that was a full time job, and one that I was peculiarly suited to and should like to undertake. I did not hanker for the administrative detail, which is not my bag."

A slightly portly man with prominent eyes, shy in a formal way but impeccably alert, Kuesel speaks in complete, well-organized paragraphs veined with irony and wit. Possibly the most literary of the partners since Parsons, he is a prolific writer, with over forty professional and technical articles to his credit, many of them unusually fresh and entertaining, and free of the ball-and-chain of technical jargon that can make convicts of even inspired ideas.

A graduate of Yale, where he went to study physics but "drifted sideways into civil engineering when the entire physics department was drafted for the Manhattan Project," he was taught by the famed Hardy Cross, whose own writings are distinguished by a captivating literary elegance and unusual moral force.

In 1947 Kuesel came to New York with three letters of recommendation to three of Cross' former students: Quade, Clint Hanover of Hardesty & Hanover and Robert Abbett of TAMS. Although Quade offered the lowest salary of the three, Kuesel liked his style.

Quade took Kuesel under his wing, but Alfred Hedefine, "who thought the world of him," became his principal sponsor and guide. Subsequently, "Michael Fiore became my professional father in bridges and Bickel in tunnels." However, for the first fifteen years of his career he was a bridge designer almost exclusively (except for a stint on Fort Ritchie), and between 1947 and 1962 most of the bridges mentioned in this book received some contribution from his hand. Then in 1963 he went to San Francisco to work on BART, where the principal problems were tunnels, "about which I knew next to nothing at all. But I fell into the hole and haven't climbed out since." The hole was vast: in the past twenty years he has participated in over 90 tunnel projects on five continents and recently edited with Bickel the massive and authoritative *Tunnel Engineering Handbook,* published by Van Nostrand Reinhold in 1982.

Kuesel is the first engineer in the firm's history (since Parsons sat in his office alone) to be its leading expert in both bridges and tunnels at the same time.

The role he chose for himself in the corporation would seem to have been just right. Variously described by his colleagues as a "genius," "the best problem-solver in the country," and "the engineer's engineer," his professional standing is perhaps best summed up by Greenfield:

> He can adapt himself to almost any problem. He seems to know everything that has ever happened in bridges and tunnels. He has a tremendous

memory. He's an enormous asset to the company, and is so recognized by his peers. Nor is he difficult to get along with—not at all. I think he's cut out of the same block as Quade: conservative in many ways, and with respect to new business opportunities tends to find more reasons not to go after something than to go after it. But once it's done, he puts his back into it. I can't say enough about Kuesel. He's just a brilliant engineer.

Of all the principals Kuesel has perhaps the most sensitive appreciation of and feeling for the company's history. Apt to refer often to "the great tradition of engineering excellence which has kept this firm going for a hundred years," he sees himself as its custodian: "the line runs with offshoots fairly directly from General Parsons through Walter J. Douglas to Eugene Macdonald to Maurice Quade to Alfred Hedefine and down to me. We have each been charged in our time with the responsibility to uphold the professional reputation of the firm, which I have taken very seriously."

CHAPTER 27

T
he corporate revolution notwithstanding, there was no abate-
ment of toil in the fields. Regional transit was a deliberate
priority, and well before BART was completed in San Francisco,
Parsons Brinckerhoff-Tudor-Bechtel (PBTB) regrouped to work on
a comparable system (known as MARTA) for Atlanta, Georgia.
Winfield O. Salter, a wry, laconic sort of man who had directed the
preliminary effort since 1966 from San Francisco, was dispatched
in 1973 to Atlanta to reorganize and expand the joint venture staff
after financing for the project was assured. He built up the staff
remarkably, from a core of 20 to about 600—largely, he says, "out of
misfits: they were good people, but they were either out of a job, or
about to be laid off, or had personality problems, or just wanted to
change locales."

A graduate of M.I.T. where he was trained as both an aeronautical
and transportation engineer, Salter's career was interrupted briefly by
a stint in the Marines during the Korean War, after which he was hired
by Bickel for the highway department under Douglas Sayers. He
excelled. Within a few years he was deputy chief highway engineer,
had founded the firm's computer department,* and not incidentally
had earned his M.S. degree at Columbia University by commuting at

---

*The computer has profoundly affected many aspects of contemporary life, and in the engineer-
ing profession, where its use has proliferated, few engineers deny its convenience as a tool.
Nevertheless, some younger engineers tend to accept what it churns out as gospel, which has led
certain of their more seasoned colleagues to warn against regarding its mathematical wizardry
with abject awe. In a recent technical paper, Kuesel observed:

"The collapse of two prominent roof structures which the firm was asked to investigate,
illustrates the dangers of computer mesmerism. In the first case, the design was supported by
an impressive computer analysis, but no one noticed that the structure analyzed by the
computer was not the one that was built.
In the second case, the computer analysis was even more impressive, but the designers
failed to follow Hardy Cross' first commandment— 'Draw the deflected structure!' When we
did, we discovered that under unsymmetrical loading the structure lurched like a circus
trapeze, and the resulting secondary stresses (which the computer did not analyze) became
the critical ones. . . Computers help in answering questions accurately. They are no help in
finding the right questions."

night to Morningside Heights. In 1959, Salter accompanied Everson west to San Francisco to labor on BART where, though he had no mass transit background, he was given special responsibility for design and route location for San Francisco, the Peninsula, and downtown Oakland. That assignment lasted seven years.

In MARTA, Salter had what amounted to a large, semi-independent company by itself.

The MARTA system is a combination rapid rail-busway that may eventually include 53 miles of track, 39 stations, and cost $3 billion. Most of the subway portion is cut-and-cover, with the notable exception of deep-level Peachtree Center Station, which was mined from below. Peachtree Street, of course, is Atlanta's most famous thoroughfare, known to millions from *Gone With the Wind.* Of more interest to the engineers, however, it rests on an unusual geologic ridge or knob of granite which divides and defines the watersheds emptying into the Atlantic and the Gulf of Mexico. This rock proved strong enough to support the carving out of three large intersecting caverns without the customary reinforcement of a thick concrete lining encasing steel ribs.

The only real precedents for exposed rock subway stations were in Stockholm, where the geology and the quality of rock are comparable. Curiously enough, new geological theories postulate that the two rock zones may in fact be linked, "that North America and Europe were once joined together and a single mountain range existed that extended from Stockholm—passing through what is now Norway, Scotland, Iceland, Newfoundland, and Labrador—all the way down the Appalachian chain to Atlanta."

During the early development of MARTA, the problems BART was having became widely known and "our Atlanta client," says Salter, "became defensive about hiring us. We had to say, 'No, we're not doing anything like BART, so don't worry about it,' and so, in general, we opted for more conservative technology."

In many respects, however, MARTA is a lot like BART. BART is fully automated in theory; both are substantially automated in fact. Both have "full train protection." The difference is that the operator on MARTA does a little more. He must start and stop the train. He must get up and operate a door switch. He's told continually what maximum speed he may operate at, and he may choose to accelerate automatically at a fixed rate or manually at a rate of his selection. In both, wayside controls reduce the speed if necessary and central computers monitor the movement of all trains. Both have unmanned automatic fare collection, electronically controlled fare gates, and round-the-clock closed-circuit T.V. surveillance in the stations. MARTA's cars are also fast, smooth, quiet, long, and spacious, with air-conditioning, carpeting, and upholstered seats.

§

The MARTA project went through some unusual changes. Though the original PBTB contract was negotiated in 1972, the Metropolitan Atlanta Rapid Transit Authority insisted in 1975 that it be renegotiated, and eventually Bechtel withdrew on the grounds that it could not accept the new compensation terms. Not without risk, Parsons Brinckerhoff and Tudor decided to shoulder the Bechtel portion alone and the new, two-firm joint venture, PB/T, continued under a new contract as of April 1, 1976.

At about this time, Salter relinquished direction of the project to his deputy, James L. Lammie, and assumed charge of the firm's new regional office in Atlanta, which he had established earlier with Arthur G. Bendelius. (Currently, Salter is the firm's technical director for all transit work.)

Lammie is a military man. He graduated from West Point with a B.S. degree in engineering in 1953, and holds three masters degrees: in engineering from Purdue, in business administration from George Washington University, and in military art from the Command and General Staff College of the U.S. Army. He spent twenty-one years in the Army in a variety of assignments, including a tour of duty in Vietnam, and later served as a staff officer in the Pentagon, and on a special review panel for the reorganization of the Army under the Army's Chief of Staff.

In the early 1970s he was appointed District Engineer for the Corps in San Francisco, where he worked on harbor development, water resources, and the construction of postal facilities; and in 1974, upon his retirement from military service, "ended up as a lobbyist," he recalls,

in the state legislature of California for a number of permit activities I'd been involved in as District Engineer. I felt there was a potential for a conflict of interest and became very uncomfortable. One day I bumped into Paul Gilbert, an old friend, and he said, 'How do you like what you're doing?' I said, 'I really don't like it.' He said, 'Oh, we've got some things going.' I said, 'Where?' He said, 'Singapore, Hong Kong, and Atlanta,' I said, 'What do you have in Atlanta?' A few days later I had dinner with Everson, Gilbert, and Michel, and ended up going to Atlanta more or less blind.

Lammie, in fact, had no transit experience in his background. "On my first trip back to the west coast," he remarked, "I rode BART so I could say I was familiar with it."

Under Lammie, construction on MARTA peaked. There were a great many people to brief, a great many plans to review. With construction placement of roughly one million dollars a day, crisis was chronic. "At one point we knocked out the power in 25 percent of Atlanta," says Lammie; on another occasion the columns being used to underpin the Merchandise Mart went into "bursting failure." Lammie met with MARTA's new assistant general manager and said, "I think we have a problem. Our building is falling down."

The building survived; with luck, the system's cars, manufactured by Société Franco-Belge also weathered the biggest bankruptcy in the history of France. "We paid for them without title," says Lammie. "They were the only items in the plant that didn't go to satisfy the claims of creditors."

"But it was fun," Lammie adds. "You go out and watch a project grow." Grow it did. In June of 1979, seven stations on the initial east-west line began operation. By 1982 over sixteen miles of track and 20 stations were serving Atlanta's eastern and western suburbs, and the mid- and downtown business zones. By 1982, however, Lammie had turned the reins over to John R. McDonald, and after a brief term as regional manager for the firm's North Atlantic Region was made head of Parsons Brinckerhoff Quade & Douglas, Inc., the firm's domestic company and major operating arm. He sits now in the red-carpeted executive suite at Penn Plaza where "you see a project maybe two or three times and you see it just as you drive by . . . you don't really get a feel for it. It's a major change."

In Pittsburgh, where the firm also studied plans for a heavy rail system in the 1960s, the city opted instead for the Westinghouse

Transit Expressway or "Skybus," an automated guideway system. The latter, a home-grown invention, was favored by the business community and Allegheny County, in a bid to make Pittsburgh the "Detroit of rapid transit." However, the firm concluded in an evaluation study it was hired to make that while the Skybus would cost less to build, it would cost decidedly more to operate and maintain—which would make it impractical. Despite Douglas' indignation, the report was suppressed, and another consultant was hired for the Skybus design. Somewhat later, feuding between the City and the County cut the work short, and in 1977 the firm returned, with Mandel as project manager and Salter as principal-in-charge, in a joint venture with Gibbs & Hill to rehabilitate Pittsburgh's trolley network into a modern light rail system.

When complete, it will cover 22.5 miles, with a new subway downtown and renovated sections to the South Hills suburbs.

§

Another engineer to emerge in the late 1960's was Martin Rubin, project manager for the track work on BART. Asked in 1966 by Everson to manage a highway project in Hawaii, Rubin soon reported back: "The opportunities here are immense." He asked for latitude to develop them, and thus founded Parsons Brinckerhoff's Hawaiian outpost. Though some of the partners expected the venture to fail, Rubin (who describes himself as "originally a technical guy") skillfully promoted the office as a local firm "which just happened to have this extra resource of talent in New York. I explained that if a special problem arose, I could just get on the phone and bring the people out here. I offered the best of both worlds."

Rubin wined and dined and talked to the "right people," developed some of the hoped-for community rapport—and the work came in: particularly, two complex highway interchanges, and the planning and design of a six-lane, ten-mile highway (known as H-3) through the volcanic Koolau Range.

Recently, upon his reassignment to southern California, Rubin was succeeded in Hawaii by Stanley Kawaguchi, who Rubin says, "has taken the best from me, discarded the bad stuff, and put his own good stuff in. Hawaii will be there forever."

§

As one might expect, BART had opened up new opportunities in California—though at first they were apparently hard to track: in 1973 Michel and Everson barely defeated a move to close the San Francisco office down. On the eve of his retirement, however, Douglas was prevailed upon to turn the office over to Gilbert, who "with the help of a number of old timers" successfully turned things around. By 1976, Gilbert's growing staff was at work on two sizable projects: PEP and SWOOP.

PEP (the Positron-Electron Project) is a 7,200-foot-long doughnut shaped tunnel for physics research at the Stanford Linear Accelerator Center (SLAC). Electrons accelerated in SLAC's two-mile-long linear accelerator are fed into a magnetic raceway and whirl clockwise around a storage ring in the PEP tunnel at close to the speed of light. Positrons are then fed into a parallel raceway in a counterclockwise direction. When the two beams of oppositely charged particles are brought together by adjusting the magnets, the high energy collisions explode the particles to reveal secrets of the structure of elementary particle physics.

PEP was Stanford's third colliding beam storage ring. It produced higher collision energies than had been previously achieved, and at the time of its completion was the cutting edge of world high energy physics research. The firm designed and managed the construction of the tunnel, as well as various half-buried experimental halls and surface support facilities.

SWOOP (or the San Francisco Southwest Ocean Outfall Project) is part of the largest public works project in the city's history. A major overhaul of the city's sewerage system, its principal feature is a 4.5-mile-long, 12-foot-diameter sewer conduit extending across the San Andreas Fault far out into the Pacific Ocean. Like the Trans-Bay Tube, the conduit is secured against earthquake rupture by a segmented design with telescoping joints.

Gilbert next began to stake some claims in southern California. In 1975 he hired Mike Schneider, opened an office in Santa Ana, and before long studies for a "multi-modal transportation center" in Orange County and an "elevated people mover" system near Disneyland were followed, in 1976, by a one-quarter million dollar transportation study for Orange County, which established a blueprint for county-wide highway and mass transportation improvements over the next twenty years. It also led to a substantial contract for the conceptual design of the 40-mile transit system the study had recommended. It was for this—and to take charge of the Santa Ana office—that Rubin was summoned from Hawaii. Under Rubin, the office has been involved in every major transit development in the Los Angeles area—including preliminary studies for a light rail network from Long Beach to Los Angeles; as a consultant to Los Angeles Metro Rail; in the preparation of a public works study to finance $14 billion worth of freeway and rail transit construction; and in miscellaneous Orange County work.

In addition to two joint venture offices in Los Angeles alone, other Parsons Brinckerhoff outposts now dot the west: in Denver, Colorado; Tempe, Arizona; and Albuquerque, New Mexico.

§

Along with the firm's adept response to the resurgent interest in mass transit, it has also figured prominently in the nation's search for sources of alternative energy—and especially for the strategic conservation of what it now has. In 1975, in order to reduce American vulnerability to an Arab oil embargo, Congress authorized creation of a strategic petroleum reserve for storing up to one billion barrels of crude oil—an amount sufficient to supply the United States for six months. In joint venture with Kavernen Bau-und-Betriebs-GmbH of West Germany the firm prepared designs for five salt dome repository sites in Texas and Louisiana. So large are some of the domes that the least of them could easily accommodate the Empire State Building.*

Quite as urgent has been the need to devise safe and effective means of storing nuclear waste. In 1971, the firm provided technical advice for the first serious study of a geologic nuclear waste storage facility in the United States, under the Savannah River Plant of the Atomic Energy Commission. However, with the best storage medium

---

*Salt caverns, whether for the production of brine or for the creation of underground storage, are created by a process called solution mining. A well is drilled into the salt dome and two concentric pipes are lowered into the completed well. Fresh water, pumped in through the inner pipe, dissolves the salt to form brine, which flows through the space between the inner and the outer pipes to the surface. As this process continues, a brine-filled cavern grows in the salt dome. Crude oil can be stored in the cavern by pumping the oil in, forcing out the brine. To retrieve the oil, the filling process is reversed.

yet to be found, the challenge has been to identify close-grained materials deep beneath the earth's surface out of which absolutely leakage-free caverns can be mined. Since 1976, in cooperation with the government's Office of Waste Isolation, the firm has studied the storage possibilities of salt, granite, shale, basalt, and tuff.

CHAPTER 28

Today, no less than in the past, the firm continues to demonstrate its time-honored strengths in tunnels, bridges, and highways. A second Hampton Roads Bridge Tunnel was begun in 1972 (directly over the site of the historic Civil War battle between the Monitor and the Merrimack) and a third Elizabeth River tunnel between Norfolk and Portsmouth in 1982. (The First or Downtown Tunnel was completed in 1952, the Midtown tunnel in 1961.)

The tubes for the third tunnel, fabricated in Texas, were to be towed 1800 miles by barge across the Gulf of Mexico, around the tip of Florida and up the Atlantic Coast to Virginia. However, while one barge was under tow in the Gulf of Mexico, a fire broke out on the tug, and after a bit of discussion the crew abandoned ship—which left the tubes floating off in the direction of South America. A rescue tug was sent down from New Orleans under forced draft, picked up the crew, and eventually found the tubes before someone could claim them as salvage.

Two other tunnels now in the making completely overshadow anything else the firm has done in the field for a long time. One is the Rogers Pass rail tunnel in Canada, on which construction began in 1984. Nine miles long, it will be the longest tunnel in the Western Hemisphere, and will feature a unique ventilation system with midtunnel gates and fans and a ventilation shaft that opens out into Glacier National Park.

The other is the Fort McHenry tunnel, now well underway, which will be the largest subaqueous highway tunnel in the world. Named for the historic fort where Francis Scott Key wrote "The Star Spangled Banner" after witnessing the unsuccessful British attack on Baltimore during the War of 1812, the tunnel is an eight-lane, double-barrelled, four-bore sunken tube leviathan under Baltimore Harbor. As with Quade's Yorktown Bridge, local patriotic feeling for the site had to be respected and "the only way the tunnel could be made acceptable to the community, was to extend it in such a way that it would be out of sight and earshot of the fort. The investment in preserving tranquility at the fort was considerable, but that's what made the project go." Each of its thirty-two huge sections is therefore different, in order to shape the curve that will carry it around the Fort McHenry peninsula. A mile and a third long, and designed in joint venture with Sverdrup & Parcel (who collaborated with the firm on the 63rd Street Tunnel) it will cost $800 million, or half as much as the whole of BART, while the amount of soil dredged up for the underwater trench (3.5 million cubic yards) equals what Parsons dug up in the dry for his IRT.

Also of recent note: for the Hood Canal in the state of Washington, the firm, in joint venture with Raymond Technical Facilities, has designed a replacement for one of the world's largest floating bridges. Destruction of the bridge was dramatic. During a heavy storm in 1979 the bridge was closed, but a semi-trailer had managed to get onto the span—whereupon, the western portion of the bridge began to break up and sink. Collapse was progressive from east to west, and as the driver backed off, the bridge disintegrated in front of him.

The western portion of the bridge and the center movable span have now been replaced with a structure two-and-a-half times as strong.

As Walter Douglas had hoped, in 1972 the challenging design assignment for Westway came through—though the project is so controversial its construction continues to be delayed.

At first glance, Westway doesn't look like much: a mere 4.2 miles of interstate route, mostly underground, from the Brooklyn-Battery Tunnel to the Lincoln Tunnel at 42nd Street along New York City's lower west side. But all agree it would permanently change the character of a historic neighborhood, including alterations in many side streets; and require a massive (230-acre) landfill in the Hudson River with possible impacts on aquatic life.

As managing engineering consultant for the project, an assignment secured in competition with 15 other firms, Parsons Brinckerhoff has responsibility for the overall engineering design.

The old West Side Highway, which Westway will replace, had been in decline for some time. Large sections had begun to buckle and split, and from Rector Street to 72nd Street the elevated portion was in partial collapse. With much of the adjacent waterfront abandoned or in disrepair, it was felt that any rehabilitation or replacement of the highway should be accompanied by redevelopment of this land.

As many as ninety members of the firm, under the supervision of Dyckman (with Art Jenny, Robert Warshaw, Bruce Podwal, and Mel Kohn as project directors) have been involved in Westway since 1972.

However, litigation has arrested its progress. Wildlife agencies have charged that the landfill would destroy 40 percent of the river's striped bass, and a Federal judge determined the preparation of the project's environmental impact statement "false and deceptive." Overall, however, opposition has been galvanized around the idea that federal funds available to help finance Westway might be traded in for equivalent funds for mass transit, road, bridge, and other urgent infrastructure repairs.

Dyckman, who clearly feels with conviction that Westway "is just under the most vicious kind of attack," confidently enumerates its benefits: it would employ 20,000 to 30,000 people, many of them from minorities, over the course of eight to ten years; clean up the west side waterfront degraded by dilapidated piers that are also a fire hazard; and provide at least 70 to 80 acres of new park land—"in effect, give the river back to the people"—as well as new land for real estate development under joint management of the city and state. Dyckman also feels Westway would be of environmental benefit: it would "actually improve the air quality," producing just one-eighth the pollution of a comparable road in the open air. "People," he says, "are genuinely trying to find out just what you can do to protect the environment, or even improve it, and at the same time feed and clothe the people and provide them with jobs. . . But the Sierra Club which has attacked the project is not so much environmentally concerned as preservation concerned. There have been more species that have disappeared from the world than exist on it at the present time. I think preservation for its own sake is a lost cause. But I wouldn't give way to anybody with regard to my concern for the reasonable protection of the environment."

"One problem," suggests Henry Michel, is that "the project has been wrongly described—it's not a highway project, it's an urban renewal project. One of its features is a highway. Every city except New York is paying close attention to the waterfront as one of its great resources. Westway creates new land, which can be redeveloped for whatever purpose. It will reduce air pollution and clean up a blighted environment."

Both Michel and Dyckman are reasonably optimistic the project will prevail. Kuesel hedges his bets:

Not having been personally involved in the thick of Westway, I'm able to take a slightly more long-range view. And my view is that people around the world regard Paris as one of the most beautiful cities in the world, which it is, because 100 years ago Baron Haussmann came through and ripped the heart out of Paris and built beautiful boulevards. One hundred years later people came to admire the Baron's beautiful boulevards, but history does not record the opinions of all the people the Baron threw out. Now we are presented with an opportunity to rebuild a substantial part of New York City, and from a 100-year point of view it's a very desirable thing. But times have changed: you have to give attention to those who are being thrown out. And where do you draw your balance? So I have no crystal ball as to where the balance will finally fall. Obviously, it is necessary to consider both the long range and the immediate view.

CHAPTER 29

During the past two decades, but especially in the last ten years, domestic expansion has been accompanied—if not paralled by—international growth. To begin with, throughout the 1960s and 1970s, led by the efforts of William Bruce, Fred Sawyer, and James Lange, the firm continued its activity in South America, with a contract for a three-state highway study in Brazil, including a belt highway for Sao Paulo, in joint venture with Edwards and Kelcey and financed by the World Bank. With assistance from Ronald Pilbeam in the field and from Dyckman, the firm's International office also undertook miscellaneous highway, airport, and ports work in Argentina, Colombia, Guyana, and Trinidad, as well as in Brazil. Another mainstay of the firm's recent South American effort has been a major role in the metro in Caracas, Venezuela. In 1966, Walter Douglas had told the Minister of Public Works: "If you don't develop a rapid transit system, the people will be walking to work—on the tops of their cars." In January 1983, the first 4.5 mile, main east-west line (a fragment of the projected 34.7-mile network of track) opened. Though less automated than MARTA, the Metro is sleek, thoroughly modern, and unusually colorful with displays of Venezuelan art in several stations.

Nevertheless, in recent years perhaps the most prosperous work abroad has been in Asia, where in 1977 the firm opened an office in Hong Kong. Like other fledgling outposts, it was originally organized around a single project, though its growth was rather remarkably nourished by the marketing of a specific new technical expertise.

One of the challenges of BART had been to design a sound ventilation system for the subway portion; and though prior to BART this had been done by rule of thumb—provide enough fan capacity to change the air so many times an hour—no one had ever tried to rationalize the complex flow of air in an interconnected network of tunnels: in which trains acted as pistons, vent shafts supplied or exhausted air en route, and stations provided large reservoirs short circuited to the surface through passenger entrances. The usual results were less than satisfactory, as attested by the unhappiness of the subway-riding public, particularly during the long hot summer months. Accordingly, Greenfield applied the judgment of a trained mechanical engineer to a problem that civil engineers had previously fumbled, and devised an improved arrangement of fans and ducts that catered both to passenger comfort and to better control of fire emergencies, and with Norman Danziger (also a mechanical engineer) he developed a novel "push-pull" concept of emergency ventilation which led to the design of a uniquely reversible fan—subsequently a standard for many new transit systems and a retrofit for some of the old.

Nevertheless, Greenfield and Danziger recognized that BART's subway ventilation design, while improved from previous practice, was still relatively crude, and while it sufficed for San Francisco, "the air-conditioned city" (from constant cool sea breezes), it would not do for Atlanta, Caracas, and other cities whose relatively hot climates demanded chilled air systems. So with the support of the Institute for Rapid Transit (now the American Public Transit Association), Parsons Brinckerhoff formed a joint venture with DeLeuw Cather & Company and Kaiser Engineers to "advance the state-of-the-art in subway ventilation environmental control systems analysis and design." The result was a handbook definitive in the field.

Danziger was project director for the program (which took some four years to complete) with Werner Metsch as chief editor of the handbook; while Woodrow Hitchcock and William Kennedy were the two most responsible for developing the key design tool, a computer simulation model of the flow and control of air, heat and moisture in a complex network of tunnels, stations and shafts.

Now, it happened that Danziger was subsequently invited to give a paper on the subway environmental simulation (SES) system at the First International Symposium on Aerodynamics and Thermodynamics in tunnels in England, attended by representatives of 26 countries. His paper, and other published articles, such as a three part series on the design of the environmental control systems for MARTA by Metsch and Arthur Bendelius, led in 1976 to a contract to evaluate the proposed system for the Hong Kong subway, which in turn led in 1977 to the design of an adequate system for an extension of the Hong Kong Metro; and out of that (through the efforts of Hitchcock, Philip

Chien, and Rubin, the regional manager for Asia) came the office in Hong Kong.

Hitchcock, who proved to be an enterprising businessman as well as a fine engineer, recognized broad opportunities for mechanical/electrical work which no local consultant had managed to snare. With the imprimateur of Air Conditioning Designers, by Appointment to Her Majesty's Imperial Hong Kong Metro System, and by collaborating with British engineers in an area of their weakness, rather than competing against them in an area of their strength, he established a secure foothold in the Crown Colony, and soon had an office staffed with 70 locals and two expatriates.

The mechanical/electrical projects multiplied—the New World Center and Regent Hotel on the tip of Kowloon peninsula overlooking Hong Kong Harbour, the Macau Ferry Terminal complex, the Lisboa hotel in Macau, and continuing work in Metro extensions both in the New Territories and on the island of Hong Kong. One of the more impressive commissions was design for a system of outdoor escalators, with a vertical rise of 400 feet and a length of a quarter mile, to provide access to the Ocean Park amusement grounds.

From Hong Kong thence to Singapore, where building on the firm's Hong Kong reputation, Hitchcock, Rubin, and Danziger concentrated on the Singapore Mass Rapid Transit Authority. Because in Singapore design and construction contracts are usually combined, the firm linked up with the formidable Japanese conglomerate Hitachi-Mitsubishi-Toshiba, and obtained the contract for the environmental control systems for the entire Singapore Metro.

Meanwhile Kuesel had parlayed his tunnel reputation into a similar association with Ohbayashi Gumi/Okamura, with whom he worked on preparing a successful tender for the design and construction of one of the Singapore stations. Within a year (by early 1984) the new Singapore office was thriving.

Elsewhere abroad, in the Middle East, under a loan from the U.S. Agency for International Development, the firm helped repair and expand the war-torn Port of Suez in Egypt, and prepared the plans for an entire new metropolis, Sadat City, midway between Alexandria and Cairo, as part of a "national strategy for channeling the country's future development away from the Valley of the Nile."

An association with Sabbour Associates, an Egyptian firm, led to the formation of the firm's Egyptian affiliate, PB Sabbour, to take advantage of agro-industrial opportunities in the region. On the other hand, it also encouraged the firm to overlook an opportunity to participate in a lucrative contract for the design of several Israeli military airfields.

One Middle Eastern undertaking—which PB embraced—has cost the firm dearly. In 1977 the firm became a major partner in a multinational consortium for design of a $2 billion project to extend the Moroccan national railway system. The 602-mile-long extension—from Marrakesh in the north across the Atlas Mountains to Laayoun in the western Sahara to the south—will require nine major bridges or trestles, and 30 tunnels, including one 6.5 miles long under the High Atlas divide at an elevation of 4133 feet. Conceivably, the line may one day form part of a system that spans the Strait of Gibraltar to Europe.

The Moroccan Railway as it exists is essentially a mining railroad. Eighty percent of its annual freight is phosphate ore. The proposed extension would increase its length by a third, and run through several important new mining areas to oil shale deposits in the south. It might also promote agriculture, and even tourism in the south—once the war with Polisario guerrillas in the western Sahara ends.

But the King of Morocco has not paid his bills. He'd like to, apparently; but he can't. The country is in dismal economic straits. The ongoing war in the south for control of the western Sahara has been costly; there has been a resulting drop in tourism; the price of phosphate, the country's principal export, has been depressed for the last five years because of world oversupply; unemployment is high; the country has the highest population growth rate in the world; and there have been two successive years of heavy drought, which means Morocco now imports more food than it exports—for the first time in its history. Even more critically, Morocco is the only North African country that has no oil. "They have no foreign exchange at all," says Michel. "All of it, in Eurodollars, goes towards buying oil."

So that is part of the problem. The other part has to do with the contract itself which was a mistake. It provided for the imposition of crippling fines without clear conditions covering their application. This, coupled with an ill-defined scope of services, allowed the client to impose requirements for work pretty much at will. In addition, the task was in any case grossly underestimated. The result is a project whose revenues are dwarfed by its costs.

Nevertheless, the firm has chosen to honor the contract.

CHAPTER 30

C orporations, to take a phrase from Robert Frost, are "thoroughly departmental," so it is not surprising that once a corporate direction was set, there were further elaborations to its scheme. In 1975 the firm was reorganized along geographic and technical lines. "In the days of the partnership," says Henry Michel, "there was very little cooperation or even communication across technical areas. Each partner ran his own discipline out of New York as if it were an independent firm. We weren't a multidisciplinary firm with interchange and interaction of talent; we were a series of single-disciplinary firms with a good deal of waste and duplication of effort. In order to diversify and grow, we had to decentralize and at the same time we had to improve managerial and technical communication firm-wide."

The firm created six regional offices, each run by a manager who to some extent functions like an old partner. The manager directs the regional office as if it were a local company, develops a style of getting business on the local level, and competes for jobs in the specific environment of the local area. But unlike the partners, regional managers are running multidisciplinary companies and can draw on the resources of the larger firm to compete successfully on a national scale. In other words, talent is shared across the firm; communication is facilitated by a centralized management structure in the corporate headquarters; and technical quality is monitored firm-wide by technical directors for each engineering discipline.

The firm not only expanded organizationally, it diversified professionally, and from 1975 to 1980 (a time remembered especially for its free-wheeling activity and ferment) grew from 500 to over 1000 employees. Traditional work continued but the openness to new ideas led to new endeavors. There were projects in agribusiness, environmental impact assessment, advanced technology, solid waste management and resource recovery, government-financed (Section 8) housing, solution mining, railroads, and real estate development. Perhaps the firm grew too fast: some of the new endeavors outstripped the resources at hand and many of the ventures withered on the vine. Others that survived evolved into separate operating companies somewhat at odds with the corporate structure of the firm.

Thus, in 1979, an umbrella or holding company, called Parsons Brinckerhoff Inc., was formed "to facilitate the firm's transition from a traditional public works oriented professional services partnership to a multi-national total services corporation." This ambitious reorientation, which emphasized economic opportunity as determined by an "executive and business management entity," was combined with a more specific holding company function: namely, as the holder of all the interests in each of the operating companies, to act as an investment advisor to the stockholders.

The largest and most prominent of the operating companies was the original corporation—Parsons Brinckerhoff Quade & Douglas, Inc.—responsible for the firm's traditional work in such areas as transit systems, highways, bridges, and tunnels. A second operating company, Parsons Brinckerhoff International, Inc. was built on the foundation of the international division, to provide services to overseas clients and to act as a holding company for foreign subsidiaries. The third evolved out of the old construction department—Parsons Brinckerhoff Construction Services, Inc. Like Parklap of old, it drew on the engineering expertise of the firm to provide construction management services and, with its own corporate structure, could pursue contracting and turnkey projects without risking the assets of the firm as a whole.

Another offspring of the holding company structure was PB-KBB, Inc., formed as an outgrowth of the salt dome petroleum reserve project for the U.S. Department of Energy (DOE). This corporation has continued to work on solution mining projects for the DOE's Strategic Petroleum Reserve Program, and in 1982 was awarded a $4.2-million contract to design the exploratory shaft facility in salt for the Office of Nuclear Waste Isolation.

The entrepreneurial approach to real estate development—to retain an equity or ownership interest in a building or facility—led to the creation of Parsons Brinckerhoff Development Corporation, Inc. Greenfield explains: "Since the firm had in-house most of the skills necessary to plan, design, and construct diverse types of facilities, with this expanded concept of project development the company now could do for itself what it used to do primarily for others." Accordingly, the firm has invested in commercial real estate and small hydropower plants with a view toward renovation.

Now, each subsidiary operating company has its own president and board of directors, elected by the board of directors of the holding company. The holding company directors, in turn, are elected by the stockholders.

While in the partnership days, ownership of stock was limited to the partners, since 1975 employees have been increasingly allowed to buy in. In Michel's estimation this is the single most significant change in the firm's recent history. It dilutes individual power; gives key employees a "piece of the rock"; provides an orderly transfer of ownership without significant impact on corporate resources, and a larger reservoir of future leadership; and, above all, is the best defense strategy against acquisition. "With this innovation," says Michel, "we have created a model for private ownership of professional service companies."

However, as the corporate structure grew, so did reservations about it. "Only good can come from frankness between colleagues," said Walter S. Douglas, whose optimistic declaration has been lately put to the test.

Both Dyckman and Lammie have described the recent changes in the firm as "traumatic," and that trauma has been felt in the extremities and joints. While there is no clinical consensus as to what this means, and while almost everyone concedes that some sort of corporate reorganization was necessary, a significant number of the firm's leading professionals are disquieted. "The organization expanded too rapidly," says a highway engineer, "the department heads are hung up on reports and paper work; they are more administrators than active in design." Milton Shedd, the firm's leading designer of movable bridges, goes further: "There is less engineering today and more management. The firm is more like a business than a profession." Joseph Goldbloom, who came to the firm in 1953 as a field engineer, concurs: "Since the company was expanded into a business corporation, I think quality has been pushed into the background as economics has taken over . . . We're top-heavy with accountants, business-getters, and everything dealing with a corporate structure." Lou Silano, the current technical director for major structures, suggests one reason: "People want to get away from the technical areas because the rewards within the company don't come from that path." Everson agrees: "Part of the professional glamour today is overshadowed by the 'numbers' and 'busy-ness.' In the old days a professional goal was to become a squad boss. Today people feel the path to success is to become a manager."

The issue, notes Silano, is not peculiar to Parsons Brinckerhoff, but can be found throughout the profession as a whole. And the firm has begun to address it in a serious way. At the prompting of Paul Gilbert, it established a career ladder for technical professionals "as an alternate route to the top." Among consulting firms the program is believed to be without precedent. "Perhaps," says Michel, "a bright young man wants to be the best sunken tube designer in the world. Maybe management *per se* is not his primary interest. He should not have to move into a management position to advance. At Parsons Brinckerhoff he will have a choice." Says another: "We are reemphasizing technical accomplishment as an end in itself, and a very desirable one." There are three rungs to the ladder. The first is Professional Associate, for those who possess recognized expertise in some field; next, Senior Professional Associate, for those with a standing beyond the firm; and, at the topmost rung, Principal Professional Associate, for those with a national or international reputation. The equivalent corporate position for the last is Vice President. Occasionally, both titles are held at once.

The Career Development Program, without quite conquering the problem, appears to have put a brake on the deterioration of professional morale.

However, the overall debate in the firm on how much emphasis to place on margins of profit as against technical excellence is not yet over. "The idea of the corporation wasn't bad," maintains Greenfield, "it just seemed to have gotten out of hand. I think we lost sight of the fact that even though we're a business—a profit-making business— we're still a profession. Too much time is being spent managing managers. We've been knocked unconscious with management and numbers."

Both Kuesel and Michel freely acknowledge that their views on the matter are diametrically opposed. According to Michel:

Kuesel and I have a differing philosophy about the engineering business. His philosophy is that the only element of any value is the technical excellence of the product. My attitude is more towards the commercial side—not to the exclusion of technical excellence—but a greater balance. And we have to be willing to sacrifice some technical excellence to maintain it . . . to be reflective of the client's needs and the client's affordability.

Kuesel, however, demurs: "Michel looks on the firm as a business enterprise whose business happens to be engineering, while I look at it as a professional services firm that incidentally must be run as a business." Differences notwithstanding, both agree that the core of the firm's business ought to be in the continuum of its experience if there is any value to having a hundred-year-old organization.

CHAPTER 31

"**O**f all human activities," ventured William Barclay Parsons, Jr. more than half a century ago, "engineering is the one that enters most into our lives, that gives us our means of living, and permeates every fiber of the social fabric." Some of the ways in which this bold pronouncement may be true are powerfully suggested by the work of the firm he founded, which in its scope, variety, and consequence has played a major role in the design of American life.

"Now, our first hundred years are receding into the past," reflects Michel, "and as time distances events from the present, they are polished by hindsight and new-found wisdom. We have learned more about the value of preserving our heritage and providing proper safeguards to slow down the spoilage to our environment. There will be more checks and balances, as a better-informed society becomes more involved and its concerns are handled more ably to develop a proper balance between desire and need." Michel and others single out the firm's expertise in underground construction (for the mining of chemicals and minerals, waste disposal and storage), in roads, bridges and tunnels (especially for repair of the nation's neglected infrastructure), and in rapid transit and water power and supply as principal pillars of prospective work.

But whatever the specific shape of things to come, it is not unlikely that in the future as in the past what is significant in the life of the profession will be boldly reflected in the life and example of the firm. From the railway to the nuclear age, its engineering accomplishments have included much that was triumphantly landmark as well as typical, and in its adaptation to each new generation of engineering need—in the face of the Depression, two world wars, and other crises—it has found its way through a thicket of difficulties to a certain renewed awareness of its historic aims. Those aims, which in the marriage of theory and practice have translated mathematics into durable monuments, make up a great tradition; and though the firm's current preoccupation with the evolution of its business structure, with "labor intensive investment," ownership transfer, and matters of finance, are of undoubted importance and timely indeed, in their fluctuating midst is to be planted for its unwavering worth such advice as Parsons urged upon the graduating class of engineers at Columbia University in 1923: "Remember that you are entering upon the practice of a noble profession. . . Seek to develop its spiritual side, to understand its ideals, and to maintain its standards of ethics and conduct. . . The engineer has a higher mission to perform than that of a mere technologist. He occupies a position of trust and great responsibility. . . The idea of *service to others* is the keynote."

So may it be.

TO KNOW ONE'S OWN ESTATE:
1965–1985

Walter S. Douglas succeeded Quade as the senior partner in 1965.

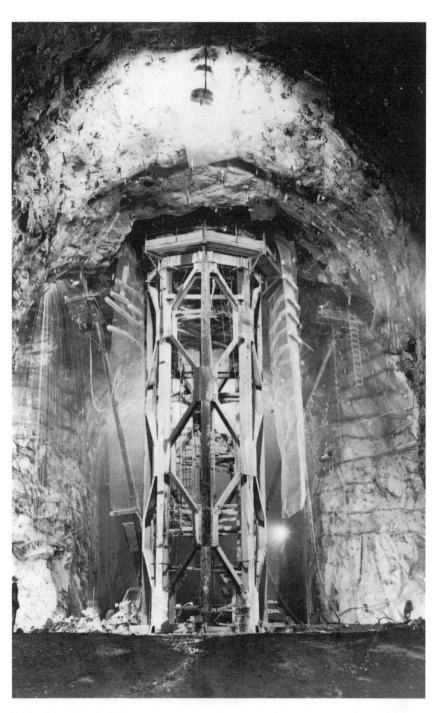

Under Douglas, the North American Air Defense Command Center (NORAD) was built—a hardened underground facility mined deep in the core of Cheyenne Mountain in Colorado and designed to resist the potential effects of nuclear weapons.

The San Francisco office for the BART project — the 75-mile-long transit system for the San Francisco Bay Area in California that set new precedents for subway construction. William Armento, who directed the cut-and-cover tunneling, in the foreground.

John E. Everson, manager of the San Francisco BART office, with some of his principal colleagues. Left to right: Martin Rubin, Thomas R. Kuesel, George J. Murphy, Everson, Winfield O. Salter.

Aerial view of Orinda Station construction site, showing the transit line emerging from the east portal of the Berkeley Hills Tunnel, 1971.

*BART was completed in 1972. Stations were hailed for their dramatic architecture
(above) and cars for their space-age design (below.)*

The 4-mile-long Trans Bay Tube linking San Francisco and Oakland, under construction, 1967.

The Newport Bridge crossing Narragansett Bay, Rhode Island.

The bridge's prefabricated, parallel wire strands were a significant advance in bridge cable construction.

Bridge designers John Swindlehurst, Louis Silano, Ahmet Gursoy at work on the Newport Bridge.

Fanfare at the opening, 1969.

Alfred Hedefine, chief designer of the Newport Bridge.

The Fremont Bridge over the Willamette River in Oregon, the longest tied-arch span in the world—another of Hedefine's major achievements.

The partners in the Douglas years (left to right): John E. Everson, Alfred Hedefine, Walter S. Douglas, William H. Bruce, Jr., Seymour S. Greenfield; (standing) Thomas R. Kuesel, Henry L. Michel, Winfield O. Salter, William T. Dyckman, Perry D. Lord.

The Halawa Interchange, one of many projects in Hawaii that resulted from the establishment of a Parsons Brinckerhoff office there.

Five Points Station, the hub of MARTA—Atlanta, Georgia's transit system—under construction.

In Peachtree Station, completed in 1982, exposed rock is used as a dramatic architectural interior finish.

Tunnel construction between Five Points and Peachtree Center stations.

Aerial structures are an integral part of the MARTA system.

A Pittsburgh trolley. In 1977 the firm began work on transforming the city's 22.5 mile long line into a modern light rail transit (LRT) system.

Construction for the LRT in downtown Pittsburgh.

Walter S. Douglas (left) and Henry L. Michel (right) at work on corporate reorganization, ca. 1974.

A tunnel boring machine works on the tunnel ring for PEP, the Positron Electron Project at Stanford Linear Accelerator Center, one of the San Francisco office's major projects of the late 1970s.

SWOOP (Southwest Ocean Outfall Project) pipe in position for launching. The 4.5 mile long, 12-foot diameter sewer conduit, part of the city's largest public works project, extends across the San Andreas Fault far out into the Pacific Ocean.

In 1978 the firm began designs for the U.S. Department of Energy's Strategic Petroleum Reserve Program, which provides for underground storage of up to 750 million barrels of crude oil in huge underground caverns along the U.S. Gulf Coast.

The formidable long-term task of design and engineering management of the Westway project, the rebuilding of New York City's West Side Highway, includes coordinating the efforts of over 30 section design consultants.

The Fort McHenry Tunnel, the largest subaqueous highway tunnel in the world, under construction 1983.

Highrise buildings in Hong Kong, one of many projects for which the firm provided mechanical/electrical design.

Townhouses in Sadat City, Egypt. The firm contributed to the master plan for this completely new metropolis 95 km. northwest of Cairo in the desert.

Pier for the new Sunshine Skyway Bridge under construction. Parsons Brinckerhoff Construction Services is in overall charge of this 15 mile long Tampa Bay crossing.

In Savannah, Georgia the firm's development corporation is converting a 19th century trainshed into a history-oriented tourist attraction — The Great Savannah Exposition.

The exterior of the trainshed has been restored; the interior will house two theaters and a museum devoted to the history of Savannah.

Henry L. Michel, President and Chief Executive Officer, Parsons Brinckerhoff, Inc., 1984.

James L. Lammie, President, Parsons Brinckerhoff Quade and Douglas, Inc., 1984.

Seymour S. Greenfield, Chairman of the Board, Parsons Brinckerhoff, Inc., 1984.

Thomas R. Kuesel, Chairman of the Board, Parsons Brinckerhoff Quade and Douglas, Inc., 1984.

The Parsons Brinckerhoff, Inc. Board of Directors 1984, (seated from left to right): Albert R. Marschall, Jose R. Bejarano, Martin Rubin, Henry L. Michel, Seymour S. Greenfield; (second row): John E. Everson, William H. Lathrop, Winfield O. Salter, Norman H. Danziger, William H. Bruce, Jr.; (standing third row): James L. Lammie, Richard Duttenhoeffer, Paul H. Gilbert, William T. Dyckman, Thomas R. Kuesel, Gen. W. T. Meredith, Ralph S. Stillman.

FOOTNOTES

Footnotes are supplied only for those sources, published and unpublished, available outside the archives of the Firm. Other quotations are from oral history interviews, articles in *Notes* (see Bibliography), or other unpublished archive material.

The Bibliography is complete.

Chapter 1
3. "drama of a single individual": Rolt, *Victorian Engineering,* p. 240.
3. "most exceptional": Butler, Introduction to Parsons, *Engineers and Engineering in the Renaissance,* p. 3.
4. "making a disturbance . . . first infringement": Parsons, *College Scrapbooks.*
6. "It is not enough": "Argument of Parsons at Governor's Hearing on Arcade Railway Bill," Albany, 23rd April, 1886.
7. "I was 35 years old": Quoted in Walker, *Fifty Years of Rapid Transit,* p. 188.
8. "against all socialistic tendencies": Letter to Shepard, February 26, 1899.
8. "the city would pay": Bobrick, *Labyrinths of Iron,* p. 223.
8. "the extent of credit": Ibid., p. 223.
8. "the laws of attraction and repulsion": Parsons, *Relativity in Thought.*

Chapter 2
10. "confidently informed he would be killed": Letter to Shepard, November 14, 1904.
11. "Although in the building": Quoted in Finch, *Engineering and Western Civilization,* p. 102.
11. "The world's progress": Parsons, *An American Engineer in China,* p. 314-15.
12. "a gay freebooter": Ibid., p. 56.
12. "pushing ahead with indomitable will": *World's Work,* Vol. 6, May 1903, p. 3469.
12. "I was the first foreigner": Parsons, op. cit., p. 56.
12. "Our actions, our songs": Ibid., p. 71.

13. "At the time of passing through his jurisdiction": Ibid., p. 78-80.
14. "Had the local cartographer": Ibid., p. 125.
14. "A nation that has had an organization": Ibid., p. 166-7.
14. "good engineering design . . . essentially American": Ibid., p. 198.
15. "We had shown no fear": Ibid., p. 50.
15. "by a singular coincidence at noon": Letter to Shepard, February 26, 1899.
15. "it took small flights of fancy": Parsons, op. cit., p. 51.
15. "a little help": Ibid., p. 313.
16. "Russia acquired a monopoly": *Encyclopaedia Brittanica,* 1965, Vol. 5, p. 586.

Chapter 3
17. "foolishly obstinate": Letter to Butler, November 14, 1904.
18. "deposited the load": New York *Times,* March 24, 1900.
18. "try to arrange": Quoted in Katz, "The New York Rapid Transit Decision of 1900," p. 91.
19. "what precautions are necessary": Quoted in Rolt, op. cit., p. 241.
20. "blushed like a schoolboy": New York *Times,* October 28, 1904.
21. "I own the bed": New York *World,* January 26, 1906.

Chapter 5
26. "the last ships . . . a mariner's nightmare": Farson, *The Cape Cod Canal,* p. 6.
27. "If ever a strip of land": Quoted in ibid., p. 5.
27. "the bare and bended arm": Quoted in ibid., p. 4.
28. "to scoop off": Ibid., p. 39.
28. "he designed a channel": Ibid., p. 32.
28. "divers had to be called in": Ibid., p. 42.
28. "If one of them lay ill": Quoted in ibid., p. 35.
29. "The first bit of rock": Quoted in ibid., p. 38.
29. "and the high tide from Cape Cod Bay": Ibid., p. 47.
30. "a barrel would float through": Letter to Belmont, January 7, 1916.
30. "knifed through a piece of colored bunting": Farson, op. cit., p. 3.
30. "with paddle wheels thrashing in reverse": Ibid., p. 51.

Chapter 7
36. "as a first contribution": Bobrick, op. cit., p. 264.
36. "Not only were the men to be sent": Parsons, *The American Engineers in France,* p. 19.
37. "the very first": Bobrick, op. cit., p. 265.

37. "From the Atlantic": Ibid., p. 265.
37. "one continuous struggle": Parsons, Letter to his mother, October 10, 1918.
38. "There was a time": Parsons, *The American Engineers in France,* p. 3-4.
39. "lay buried under the ruins": Ibid., p. 5.
39. "Colonel is like father in a family": Letter to Butler, June 25, 1922.

Chapter 9
54. "as a real contribution": Butler, Introduction to Parsons, *Engineers and Engineering in the Renaissance,* p. 3.
54. "He was a true representative": Quoted in New York *Times* obituary of Parsons, May 10, 1932.
55. "General Parsons had certain qualities": Quoted in *National Cyclopaedia of American Biography,* p. 1001.

Chapter 22
174. "once launched": Bobrick, op. cit., p. 316.

The photographs in the four portfolios come from the archives of Parsons Brinckerhoff with the exception of those noted below:

58. top—courtesy of Columbia University School of Engineering and Applied Science.
58. bot.—Civil Engineering Archival Collection, Division of Mechanical & Civil Engineering, National Museum of American History, Smithsonian Institution.
61. top and bot.—Civil Engineering Archival Collection, Division of Mechanical & Civil Engineering, National Museum of American History, Smithsonian Institution.
65. top—Civil Engineering Archival Collection, Division of Mechanical & Civil Engineering, National Museum of American History, Smithsonian Institution.
65. bot.—from *Proceedings of the American Society of Civil Engineers,* v. 74, no. 1210, p. 214-254 (1911).
67. bot.—Civil Engineering Archival Collection, Division of Mechanical & Civil Engineering, National Museum of American History, Smithsonian Institution.
67. top—Courtesy of Mrs. Elizabeth Brinckerhoff.
71. top and bot.—Civil Engineering Archival Collection, Division of Mechanical & Civil Engineering, National Museum of American History, Smithsonian Institution.

111. top—from *Proceedings of the American Society of Civil Engineers,* v. 106, no. 2124, p. 1391-1436 (1941).
113. bot.—Courtesy of Will Brown.
240. bot.—Barry Rokeach.
241. top—Robert Nickelsberg.
242. bot.—Robert Nickelsberg.
244. Courtesy of SKYCEI, St. Petersburg, Fla.
245. top—Hank Ramsey.
245. bot.—Jeanne Papy.
246. top—David Sailors.
247. top and bot.—David Sailors.
248./249.—David Sailors

I. THE PARSONS BRINCKERHOFF ARCHIVES

A. Transcribed Oral History Interviews

William Armento
John O. Bickel
Wilson Binger
Paul Brautigam
William H. Bruce
Tio Chen
Walter S. Douglas
William T. Dyckman
John E. Everson
Michael E. Fiore
Paul H. Gilbert
Joseph Goldbloom
Seymour S. Greenfield
Keith Hawksworth
Mrs. Julia Hedefine
Eugene Hardin
William Jones
Arthur Jenny
John Kalapos
Thomas R. Kuesel
James Lammie

James Lange
Perry Lord
Charles Louis
Gerald T. McCarthy
W. Thomas Meredith
Henry L. Michel
Chris Murphy
George J. Murphy
Fritz Panse
Martin Rubin
Mario Salvadori
Joseph Sassani
Fred Sawyer
David Seader
Milton Shedd
Lou Silano
Roger B. Stevenson
Gerald Sturman
Les Sutcliffe
Donald Tanner
Joseph Thomas

James Thoresen   David Werblin
Bert Tryon   John J. White
George Vaccaro   Helen Yasso
Robert Warshaw

B. Articles in *Notes*, the in-house publication of the firm, 1952-1982.

"Accomplishments in 75 Years of Practice," February 1960, 2-16.
"Albany Airport Developments Over the Past Fifteen Years," May 1962, 1-5.
"Alfred Hedefine, Building 'The World of Tomorrow,'" Spring 1978, centerfold.
"Alternative Energy Market Burns Bright," Fall 1980, 1-6.
"Annual Engineering Services," Spring 1965, 11-13.
"Architecture Enhances Engineering Projects," June 1959, 1-3.
"Baltimore Area Mass Transportation Study," Fall 1965, 3-16.
"Baton Rouge Sewerage System," Summer 1963, 1-6.
"Baytown Tunnel Nears Completion," July 1953, 1-2.
"The Cape May Lewes Ferry," Summer 1966, 6-10.
"A Capital Achievement," Fall 1972, 3-13.
"Caracas 'Metro' to Brighten City's Transit Picture," Summer 1968, 4-11.
"Causeway Adds Pelican Island to Galveston's Area," March 1958, 1-3.
"The Chairman of the Board," January 1957, 1-3.
"China Revisited as Firm's Ninth Decade Begins," Spring 1975, 17.
"Clearing the Decks for Intermodal Container Shipping," Fall 1968, 3-7.
"CM Team Calls the Signals for Philadelphia Center City Commuter Rail Connection," Spring 1981, 1-5.
"Cool, Clean Water: A Report," Summer 1969, 2-9.
"Development and Expansion of Peru's Biggest Port," Spring 1968, 2-9.
"Driving Through Blue Mountain," April 1957, 1-3.
"Egypt Embarks on a 25-Year Port Program," Summer 1979, 4-6.
"Elizabeth River Tunnels—A Decade of Service," Winter 1962, 1-5.
"Engineering in the Field," November 1957, 6-11.
"The English Channel Tunnel: Dream or Reality?," Winter 1981, centerfold.

"The Firm's Early History," December 1960, 1-12.

"Fort McHenry Tunnel Makes a Splash," Winter 1981, 1-6.

"From Concept to Construction: A Chronology of Key Events in the Development of the San Francisco BART Project," Winter 1963-1964, 20-21.

"The Garden State Parkway," Fall 1954, 1-3.

"The Growth of an Airport," Winter 1974-1975, 1-6.

"Halawa Interchange Complex," Winter 1968, 5-11.

"Hamilton Transportation Study," Summer 1964, 2-7.

"Highway Corridor Planning," Fall-Winter 1973, 1-5.

"An Interstate Highway Crosses Northern New Jersey," Spring 1961, 1-4.

"John E. Everson Elected Chairman of the Board," Fall 1977, 13.

"Lammie Named President of Domestic Company," Fall 1982, 11.

"The Long Island Rail Road Gets Immediate Action Program," Winter 1967, 1-10.

"Looking Back, Looking Forward," Fall 1970, 1-29.

"LRT Comes to Pittsburgh," Spring 1979, 1-5.

Macdonald, E.L., "Seven Decades of Engineering Progress," Winter 1955, 2-4.

_____, "So That Each May Know All," November 1952, 1.

"The Magnum Opus of William Barclay Parsons," Spring 1980, centerfold.

"Main Street-New York State," Spring 1965, 1-3.

"The Man Who Planned the Subway," Fall 1969, centerfold.

"Mandate for Change," Spring 1976.

"MARTA's Open: Relax and Enjoy the Ride," Fall 1979, 1-10.

"MARTA's Peachtree Center Station," Fall 1977, 2-5.

"Mattapan: A School for the Community," Summer 1974, 1-11.

"Maurice N. Quade, An Engineer's Engineer," Spring 1982, centerfold.

"Maurice N. Quade Retires as Active Partner," Summer 1966, 13-17.

"Modern Piers Improve the Shores of the Hudson," Summer 1955, 1-3.

"The Modernization of Muni," Spring 1975, 1-5.

Moses, Lisa, "Henry M. Brinckerhoff," Spring 1981, centerfold.

"A Most Complete Engineer," Spring 1977, centerfold.

"Moving Ahead on the Richmond-Petersburg Turnpike," Summer 1971, 1-9.

"National Electric Service: A Corporate Profile," Spring 1967, 13-15.

"New Approaches to Transit Facilities Architecture," Summer 1968, 12-23.

"New Jersey Opens Bergen-Passaic Expressway," Spring 1965, 3-8.

"New Program Promotes Professional Excellence," Fall 1980, 8-9.

"A New World's Record Vertical Lift Bridge," November 1959, 1-3.

"The Newport, Rhode Island Suspension Bridge: A Better Way to Cross the Bay," Fall 1969, 1-24.

"Newport, Rhode Island Suspension Bridge Nears Completion," Summer 1969, 18-19.

"NORAD Combat Operations Center," Winter 1963-1964, 2-7.

"Opportunity Park Garage: A Sophisticated Piece of Engineering," Summer 1972, 3-11.

"Over and Under Hampton Roads," November 1957, 1-5.

"Parklap Construction Companies," Spring 1979, centerfold.

"Parsons Brinckerhoff and the Panama Canal," Fall 1978, centerfold.

"Parsons Brinckerhoff Airport Consulting Services Here and Abroad," Fall 1967, 1-9.

"Parsons Brinckerhoff Operations in Latin America," Winter 1969, 3-10.

"Parsons Brinckerhoff: The Framework for Operations," Spring 1967, 3-12.

"The Parsons Plaques," Spring 1965, 14-15.

"PB Asia—Five Years of Growth in Hong Kong," Spring 1982, 1-4.

"PBDC Steers Firm into Project Development," Summer 1981, 1-5.

"PBQ&D on Baltimore Urban Design Conception," Summer 1968, 24-25.

"Pequannock Waters Secured for the City of Newark," November 1961, 1-6.

"A Plan for Ports and Waterways of Peru," Fall 1966, 20-23.

"Plant Expansion Paces Akron's Growing Water Supply," Summer 1965, 3-8.

"Rapid Transit for the San Francisco Bay Area," Fall 1956, 1-7.

"Reaching Across the Bay," Spring 1954, 1-3.

"Sadat City, New City to Rise on Egypt's Desert," Spring 1978, 2-4.

"Salt Dome Caverns Store U.S. Oil Reserve," Fall 1978, 1-7.

"San Francisco's SWOOP—Major Marine Pipeline for Clean Water Program," Fall 1982, 1-6.

"Sanitary Sewer Improvements for Hamilton Township," Summer 1967, 3-7.

"Sewage Treatment for the City of Binghamton," June 1960, 1-6.

"Shenango Dam and Reservoir," Spring 1966, 3-9.

"Short Cut at Savannah," Summer 1954, 1-3.

"Small Hydro Makes a Comeback," Spring 1980, 1-3.

"St. Louis: Case History of a Transit Study," Spring 1972, 3-12.

"Staff Notes," Summer 1968, 31-34.

"The Suburbanization of Fairfax," Winter 1977, 1-5.

"The Superhighway of the Old Dominion," August 1958, 1-3.

"Tall Tunnel for Manhattan," Spring 1974, 16-25.

"To Protest Ugliness, Create Beauty," Spring 1970.

"Toledo Port Development," Winter 1966, 3-8.

"Tomorrow's Water for Denver," Fall 1974, 1-11.

"Transbay Tube Construction," Fall 1967, 10-15.

"Tunnel Engineering Throughout the World," Fall 1966, 3-10.

"Tunneling Through the Depression," Fall 1977, centerfold.

"Tunnels for the BART Project," Spring 1970, 2-7.

"Twice as Much Water for Youngstown and Niles," January 1959, 1-3.

"Two Firms Affiliate with PBQ&D," Spring 1966, 28-29.

"Two Remarkable Architects," Fall 1980, centerfold.

"Unique Structures Designed to Drill for Offshore Oil," July 1957, 1-3.

"Urban Design Team Concept Aids Vancouver Transportation Study," Spring 1969, 12-21.

"Walter J. Douglas, Master of Masonry Bridges," Summer 1981, centerfold.

"Waterfront Engineering Around the World," Winter 1964, 3-10.

"We Are Proud to Have Served," February 1953, 1.

C. Additional Material

Bickel, John O. "Advance Base Floating Dry Docks." Address before the Power and Industrial Group of the New York Section A.I.E.E., October 10, 1945.

––––––––––. "Letter to Walter S. Douglas," December 6, 1973.

––––––––––. *Memoir.*

Brill, Elizabeth. "Biographical Memoir of My Father, Eugene Klapp," 1981.

Brown, W.K. "Letter to Rena Frankle," November 25, 1980.

Cross, Wilbur. *An Enduring Heritage. Seventy-five Years of Distinguished Service in Engineering.* New York: Parsons Brinckerhoff Quade & Douglas, 1960.

Danziger, N.H. "Confidential Memorandum on Operational Plan to Henry L. Michel," July 11, 1978.

Douglas, Walter J. "Letter of Retirement, to Partners," December 31, 1940.

Douglas, Walter S. "An Enduring Heritage: Ninety Years of Progress in Engineering, Planning and Architecture." Speech to the Newcomen Society, 1975. The Newcomen Society in North America, New York, 1975.

_____. "Educating the Engineer," n.d.

_____. "Epitome of Remarks Made Before the National Advisory Council on Career Education," November 9, 1976.

_____. "Letter to John O. Bickel," January 10, 1974.

_____. "Letter to Leland Hazard," February 17, 1976.

_____. "Letter to Martin Rubin," January 21, 1974.

_____. "Letter to Paul H. Gilbert," April 15, 1976.

_____. "Letter to Rebecca Yamin," November 19, 1982.

_____. "Letter to Roger B. Stevenson," April 19, 1977.

_____. "Letter to Stig Henrikson," November 6, 1972.

_____. "Memorandum of Profit and Loss in 1939," addressed to the Directors, June 29, 1977.

_____. "Subways." Essay, n.d.

Elvove, Eli. "Memo to Walter S. Douglas," February 18, 1977.

Gilbert, Paul H. "Comments on Draft of Operational Plan to Henry L. Michel," September 4, 1979.

Greenfield, Seymour S. "Letter to Citizens Stamp Advisory Committee," September 20, 1982.

_____. "Memorandum on Holding Company Operational Plan to Henry L. Michel," May 24, 1979.

Grosjean, P.A.W. "Letter to Seymour S. Greenfield," September 23, 1982.

Halmos, E.E., Jr. "Letter to Thomas R. Kuesel," February 18, 1980.

Hazard, Leland. "Letter to Walter S. Douglas," December 30, 1975.

Hogan, John P. "Memoir," ca. 1955.

_____. "Record of Professional Experience," August 31, 1939.

_____. "Qualifications of John P. Hogan," September 1, 1939.

_____, and Macdonald, E.L. "Confidential Memorandum

to the Partners on Business Prospects for the Coming Years," May 6, 1943.

Hutton, F.R. "Letter to William Barclay Parsons, Jr.," December 4, 1884.

Jaeger, H.F. "Confidential Memorandum on Corporate Organizational Goals to Henry L. Michel," July 15, 1975.

——————. "Letter to Henry L. Michel," June 30, 1975.

——————. "Memorandum to Directors on Cash Problems," March 26, 1965.

Krulik, Casimir. "Financial Report of the Firm for the Calendar Year 1945," submitted to Partners, February 20, 1946.

——————. "Letter to Rena Frankle," October 11, 1979.

——————. "Notes on, and Corrections to, Hogan's Memoir," addressed to Rena Frankle, September 6, 1979.

——————. "Resumes of the Activities of the Parklap Affiliates," addressed to Rena Frankle, September 6, 1979.

Kuesel, Thomas R. "Memoir of Walter S. Douglas," prepared for *Notes* centerfold, January 4, 1977.

——————. "Memorial Tribute to Alfred Hedefine (1906-1980)," prepared for the National Academy of Engineering of the United States of America, April 27, 1981.

——————. "Remarks on a draft for Quade centerfold in *Notes,*" January 1982.

——————. "Memorandum on TPF&C Report on Operational Plan to Henry L. Michel." May 9, 1979.

——————. "Memo to Henry L. Michel on Operational Plan," July 9, 1979.

——————. "Suggested Changes in Draft Statement of Corporate Philosophy," January 9, 1979.

——————. "Business Accomplishments," handwritten narrative summary, 1972.

McNulty, Nancy J. "Research Information Collected on the Occasion of the Seventy-fifth Anniversary of Parsons Brincker-hoff Quade & Douglas," in 2 vols.

Michel, Henry L. "Final Draft of Parsons Brinckerhoff Inc. Operational Plan," October 5, 1979.

Parsons, William Barclay, "Letter to E.S. Bowers," November 4, 1883.

Parsons, William Barclay, Jr. "Farewell Speech to Eleventh Engineers Regiment," April 30, 1919.

——————. "Letter to Alfred Noble," March 13, 1903.

_____. "Letter to Alfred Noble," March 26, 1903.

_____. "Letter to Alfred Noble," June 22, 1905.

_____. "Letter to August Belmont," March 25, 1903.

_____. "Letter to William T. Cosgrove," August 4, 1931.

_____. "War Letters, 1917-1919."

Penkul, H.J. "Memorandum to Henry L. Michel," July 9, 1975.

"Professional Record of Parsons Brinckerhoff Hogan & Macdonald, Engineers," October 1945.

"Professional Record of Parsons Brinckerhoff Hall & Macdonald— Water Supply and Hydraulic Works," prepared for Frank H. Backstrom, City Manager, Wichita, Kansas, May 11, 1956.

Quade, Maurice N. "Letter, re: West Rock Tunnel," October 20, 1948.

Rubin, Martin. "Confidential Memorandum on Operational Plan to Henry L. Michel," July 10, 1975.

Salter, Winfield O. "Confidential Memorandum on Nominees for RACM Representative on Holding Company to Thomas R. Kuesel," March 16, 1979.

_____. "Letter to Henry L. Michel," July 10, 1975.

_____. "Letter to Henry L. Michel," April 30, 1979.

Thoresen, Soren A. "Brief Outline of Principal Engineering Jobs from June 1, 1905 to June 1, 1949," March 15, 1949.

Washington, Booker T. "Letter to William Barclay Parsons, Jr.," April 4, 1911.

"William Barclay Parsons, Jr. as Trustee of the Carnegie Institution," prepared by the Carnegie Institution of Washington, 1982.

Yamin, Rebecca. "Annotated Bibliography of the Parsons Brinckerhoff Archives," January 1983.

II. OTHER SOURCES

A. *Special Manuscript Collections:* Columbia University, Butler Library (Nicholas Murray Butler Papers and Correspondence [NMB], Seth Low Correspondence [SL], Edward Morse Shepard Correspondence [EMS]); New York Historical Society (Belmont Papers [BP]).

Butler, Nicholas Murray. "Letter to William Barclay Parsons, Jr.," November 24, 1913 [NMB].

—————. "Letter to William Barclay Parsons, Jr.," November 20, 1924 [NMB].

—————. "Letter to William Barclay Parsons, Jr.," August 23, 1927 [NMB].

—————. "Letter to William Barclay Parsons, Jr.," January 30, 1932 [NMB].

Elliott, Howard. "Letter to August Belmont," January 7, 1916 [BP].

—————. "Letter to August Belmont," February 16, 1916 [BP].

Parsons, William Barclay, Jr. "Letter to August Belmont," July 25, 1914 [BP].

—————. "Letter to August Belmont," January 14, 1916 [BP].

—————. "Letter to August Belmont," February 8, 1916 [BP].

—————. "Letter to August Belmont," March 2, 1916 [BP].

—————. "Letter to August Belmont," November 20, 1922 [BP].

—————. "Letter to Edward Morse Shepard," February 26, 1899 [EMS].

—————. "Letter to Edward Morse Shepard," April 24, 1899 [EMS].

—————. "Letter to Edward Morse Shepard," March 24, 1901 [EMS].

—————. "Letter to Nicholas Murray Butler," November 14, 1904 [NMB].

—————. "Letter to Nicholas Murray Butler," June 25, 1922 [NMB].

—————. "Letter to Seth Low," May 9, 1900 [SL].

—————. "Letter to Seth Low," May 11, 1900 [SL].

—————. "Letter to Seth Low," December 6, 1907 [SL].

—————. "Letter to Seth Low," January 24, 1908 [SL].

B. *Articles and Books*

Alder, June. "Longboat Man to Help Build New Skyway," *The Islander,* May 6, 1982. 1, 12-13.

Bobrick, Benson. *Labyrinths of Iron, A History of the World's Subways.* New York: Newsweek Books, 1981.

Catalogue of the William Barclay Parsons Collection. Department of Science and Technology, The New York Public Library.

Ellis, William A. "Erecting a 544-ft Lift Bridge," *Engineering News-Record,* January 30, 1936.

Farson, Robert H. *The Cape Cod Canal.* Middletown, CT: Wesleyan University Press, 1977.

Finch, James Kip. *Engineering and Western Civilization.* New York: McGraw-Hill, 1951.

Foster, H. Alden and Halmos, Eugene E. "Memoir (1195) - Walter Jules Douglas," American Society of Civil Engineers, 1941.

Hedefine, Alfred and Hardesty, Shortridge. "Superstructure of Theme Building of New York's World's Fair," *Transactions,* American Society of Civil Engineers, September 1940.

"Historian Helps Corporations Recall the Story of their Past," *New York Times,* May 25, 1980.

Hoover, Thomas E. "The New York Subway From Alfred Eli Beach to William Barclay Parsons," *Mass Transit,* July/August 1976.

Katz, Wallace B. "The New York Transit Decision of 1900: Economy, Society, Politics," Historic American Engineering Record, Interborough Rapid Transit Survey, 1974.

"Lift Span over Cape Cod Canal Sets New Precedents," *Engineering News-Record,* January 30, 1936.

Macdonald, Eugene L. "Roller Bearing Design for a 544-foot Lift Span," ibid.

Mazza, Frank. "Conversation with Walter S. Douglas," *People,* February 1976, 18-20, 29.

"An Old Firm Forges Success from Versatility," *Engineering News-Record,* February 4, 1965, 48-49.

"Maurice N. Quade," *Consulting Engineer,* November 1965, 12-20.

McDonald, John B. "Tunneling New York — Showing How the Metropolis Has Solved the Rapid Transit Problem," *Colliers Weekly,* December 22, 1900.

"The Newport Bridge," *Engineering News-Record,* June 13, 1968, 42-44.

"Parsons Brinckerhoff's High Stakes Players," *Engineering News-Record,* September 30, 1982, cover story.

Parsons, William Barclay, Jr. "Address Before the Washington Society of Engineers," Washington, D.C., March 1, 1916.

_____. "Address Delivered at the Annual Service Commemorating Washington's Birthday." New York, Church of the Holy Communion, 1922.

_____. "Address at the Seventh Annual Banquet of the Boston City Club." Boston, February 8, 1916.

_____. *An American Engineer in China.* New York: McClure, Phillips & Co., 1900.

_____. *The American Engineers in France.* New York: D. Appleton & Co., 1920.

_____. "The Architect and the Engineer." An Address delivered at the Architectural League. New York, February 8, 1911.

_____. "Argument on the Arcade Railway Bill." Pamphlet, New York, 1886.

_____. "Engineering and Economics." An address delivered at the inaugural meeting of the Columbia University Chapter of the American Society of Civil Engineers. New York, November 3, 1927.

_____. "Engineering as a Profession." New York, 1902.

_____. *Engineers and Engineering in the Renaissance.* Baltimore: Williams & Wilkins, 1939.

_____. "College Scrapbooks," William Barclay Parsons Collection, New York Public Library.

_____. "IRT Scrapbook," William Barclay Parsons Collection, New York Public Library.

_____. "Manifestation of Natural Laws in Human Nature." A commencement address delivered at Trinity College, Hartford, June 21, 1921.

_____. "Progress in Engineering." An address delivered at Columbia University, New York, November 21, 1929.

_____. "Relativity in Thought." A commencement address delivered at the Carnegie Institute of Technology, June 8, 1926.

_____. *Report to the Board of Rapid Transit Commissioners in and for the City of New York on Rapid Transit in Foreign Cities.* New York, 1894.

_____. "Rapid Transit in New York," *Scribner's,* May 1900.

_____. "Skyscrapers." *Cambridge University Engineering and Aeronautical Society Journal* (1930), 23-40.

_____. *Track; A complete Manual of Maintenance of Way.* New York, 1885.

_____. *Turnouts: Exact Formulae for their Determination, Together with Practical and Accurate Tables for Use in the Field.* New York: Engineering News Publishing Co., 1884.

"Port of Havana Docks Company," *The South American,* August 1916.

Quade, Maurice N. "Special Design Features of the Yorktown Bridge," *Transactions,* American Society of Civil Engineers, January 1953, 109-23.

Ridgway, Robert and Bevan, Lynne J. "Memoir (546) - Harry de Berkeley Parsons," American Society of Civil Engineers, 1935.

Rolt, L.T.C. *Victorian Engineering.* New York: Penguin, 1970.

Scott, Charles. "Design and Construction of the IRT," *Civil Engineering,* 1978.

Smith, D. and Steadman, L.E. "Present Value of Corporate History," *Harvard Business Review,* November-December 1981.

"Specify New Circular Tubes for Texas Highway Tunnel," *Engineering News-Record,* May 19, 1949.

"Sunken Tubes Lowered into Tricky Waters," *Engineer,* June 10, 1971, 22-24.

Thoresen, Soren A. "Shield-Driven Tunnels Near Completion Under the Scheldt at Antwerp," *Engineering News-Record,* June 29, 1933, 827-32.

_____. "Tunnel Lining of Welded Steel," *The Iron Age,* April 30, 1930, 985-89.

"Tribute to Swiss Engineer, John Otto Bickel, Renowned Authority on Tunnels," *Swiss American Review,* October 20, 1982, 5-6.

Walsh, George. *Gentleman Jimmy Walker, Mayor of the Jazz Age.* New York: Praeger, 1974.

INDEX

267

Interstate Highway Program, 141
Isthmian Canal Commission, 25
Itabira (Brazil), 99-102

Jackson, Andrew, 46
Jamestown Bridge (Rhode Island), 85, 178
Jenny, Arthur, 139, 142, 183, 206
Jiménez, Marcos Pérez, 182
John E. Mathews Bridge (Florida), 130
Johnson, Stanley, 136

Kaiser Engineers (firm), 209
Kauffman, Herbert, 141
Kavernen Bau-und-Betriebs-GmbH (firm),
    202
Kawaguchi, Stanley, 200
Keflavik (Iceland), 123-25, 193
Kellog, M. W., 189
Kennedy, William, 209
Ketchum, 52
Key, Francis Scott, 204
King, Elwyn H., 168, 169
Kittredge, Henry, 27
Klapp, Eugene A., 24-25, 35, 42
    Florida construction by, 48
    during Great Depression, 50-51
    in Parklap Construction Companies,
        45-46
    during World War I, 36, 39
Knappen, Theodore, 91, 95, 99-101, 103, 122
Kohn, Mel, 142, 206
Kuesel, Thomas R., 122, 185
    BART system and, 169
    on computers, 196n
    as "conscience" of firm, 193-95
    on Walter S. Douglas, 187
    on Hedefine, 176, 180
    on NORAD headquarters, 165-66
    on philosophy of engineering business,
        216-17
    on Quade, 140
    in Singapore, 210
    on Westway, 207

Lammie, James L., 198-99, 215
Lange, James, 208
Lehigh Tunnel (Pennsylvania), 132
Leopold (king, Belgium), 10
London, subway system of, 6

Lord, Perry, 183-84
Lord & Den Hartog, Architects and
    Engineers (firm), 183
Los Angeles, 202
Louis, Charles, 142
Lulla, J. B., 210

McCarthy, Gerald T., 101-3, 104-5
Macdonald, Eugene, 82-85, 95, 98, 134
    Bruce and, 135, 136
    on firm's finances, 104, 105
    firm under, 185
    Garden State Parkway and, 128, 129
    Hall and, 136
    on Knappen, 99
    made partner, 92
    on Quade, 138
    retirement of, 136, 138
    as senior partner, 123
McDonald, John B., 17, 20, 21
McDonald, John R., 199
Macloskie, Charles, 32
Maloney, Edward, 82
Mandel, Herbert, 179, 200
marine engineering, 181-82
    during World War I, 39-40
    during World War II, 96-98
    in 1950s, 141
Marshall, Samuel W., 101, 102
MARTA rapid transit system (Atlanta), 181,
    196-99
Massachusetts
    Buzzards Bay Bridge in, 85, 91
    Cape Cod Canal in, 26-31, 40-41
McKim, Mead & White (firm), 46
Meade, Robert H., 141
Metropolitan Atlanta Rapid Transit
    Authority, 198
Metsch, Werner, 209
Michel, Henry L., 185, 189-91
    on advancement within firm, 216
    on employee stock ownership, 215
    on firm's first hundred years, 218
    on firm's structure, 187-88, 213
    on Moroccan situation, 211
    on philosophy of engineering business,
        216-17
    San Francisco office and, 201
    on Westway, 206-7